More Hunting Wasps

J. Henri Fabre

CONTENTS

CHAPTER 1.

THE POMPILI.

***

The Ammophila's caterpillar, the Bembex, gadfly, the Cerceris, Buprestis and Weevil, the Sphex, Locust, Cricket and Ephippiger: all these inoffensive peaceable victims are like the silly Sheep of our slaughterhouses; they allow themselves to be operated upon by the paralyser, submitting stupidly, without offering much resistance. The mandibles gape, the legs kick and protest, the body wriggles and twists; and that is all. They have no weapons capable of contending with the assassin's dagger. I should like to see the huntress grappling with an imposing adversary, one as crafty as herself, an expert layer of ambushes and, like her, bearing a poisoned dirk. I should like to see the bandit armed with her stiletto confronted by another bandit equally familiar with the use of that weapon. Is such a duel possible? Yes, it is quite possible and even quite common. On the one hand we have the Pompili, the protagonists who are always victorious; on the other hand we have the Spiders, the protagonists who are always overthrown.

Who that has diverted himself, however little, with the study of insects does not know the Pompili? Against old walls, at the foot of the banks beside unfrequented footpaths, in the stubble after the harvest, in the tangles of dry grass, wherever the Spider spreads her nets, who has not seen them busily at work, now running hither and thither, at random, their wings raised and quivering above their backs, now moving from place to place in flights long or short? They are hunting for a quarry which might easily turn the tables and itself prey upon the trapper lying in wait for it.

The Pompili feed their larvae solely on Spiders; and the Spiders feed on any insect, commensurate with their size, that is caught in their nets. While the first possess a sting, the second have two poisoned fangs. Often their strength is equally matched; indeed the advantage is not seldom on the Spider's side. The Wasp has her ruses of war, her cunningly premeditated strokes: the Spider has her wiles and her set traps; the first has the advantage of great rapidity of movement, while the second is able to rely upon her perfidious web; the one has a sting which contrives to penetrate the exact point to cause paralysis, the other has fangs which bite the back of the neck and deal sudden death. We find the paralyser on the one hand and the slaughterer on the other. Which of the two will become the other's prey?

If we consider only the relative strength of the adversaries, the power of their weapons, the virulence of their poisons and their different modes of action, the scale would very often be weighted in favour of the Spider. Since the Pompilus always emerges victorious from this contest, which appears to be full of peril for her, she must have a special method, of which I would fain learn the secret.

In our part of the country, the most powerful and courageous Spider-huntress is the Ringed Pompilus (Calicurgus annulatus, FAB.), clad in black and yellow. She stands high on her legs; and her wings have black tips, the rest being yellow, as though exposed to smoke, like a bloater. Her size is about that of the Hornet

(Vespa crabro). She is rare. I see three or four of her in the course of the year; and I never fail to halt in the presence of the proud insect, rapidly striding through the dust of the fields when the dog-days arrive. Its audacious air, its uncouth gait, its warlike bearing long made me suspect that to obtain its prey it had to make some impossible, terrible, unspeakable capture. And my guess was correct. By dint of waiting and watching I beheld that victim; I saw it in the huntress' mandibles. It is the Black-bellied Tarantula, the terrible Spider who slays a Carpenter-bee or a bumblebee outright with one stroke of her weapon; the Spider who kills a Sparrow or a Mole; the formidable creature whose bite would perhaps not be without danger to ourselves. Yes, this is the bill of fare which the proud Pompilus provides for her larva.

This spectacle, one of the most striking with which the Hunting Wasps have ever provided me, has as yet been offered to my eyes but once; and that was close beside my rural home, in the famous laboratory of the harmas. I can still see the intrepid poacher dragging by the leg, at the foot of a wall, the monstrous prize which she had just secured, doubtless at no great distance. At the base of the wall was a hole, an accidental chink between some of the stones. The Wasp inspected the cavern, not for the first time: she had already reconnoitred it and the premises had satisfied her. The prey, deprived of the power of movement, was waiting somewhere, I know not where; and the huntress had gone back to fetch it and store it away. It was at this moment that I met her. The Pompilus gave a last glance at the cave, removed a few small fragments of loose mortar; and with that her preparations were completed. The Lycosa was introduced, dragged along, belly upwards, by one leg. I did not interfere. Presently the Wasp reappeared on the surface and carelessly pushed in front of the hole the bits of mortar which she had just extracted from it. Then she flew away. It was all over. The egg was laid; the insect had finished for better or for worse; and I was able to proceed with my examination of the burrow and its contents.

The Pompilus has done no digging. It is really an accidental hole with spacious winding passages, the result of the mason's negligence and not of the Wasp's industry. The closing of the cavity is quite as rough and summary. A few crumbs of mortar, heaped up before the doorway, form a barricade rather than a door. A mighty hunter makes a poor architect. The Tarantula's murderess does not know how to dig a cell for her larva; she does not know how to fill up the entrance by sweeping dust into it. The first hole encountered at the foot of a wall contents her, provided that it be roomy enough; a little heap of rubbish will do for a door. Nothing could be more expeditious.

I withdraw the game from the hole. The egg is stuck to the Spider, near the beginning of the belly. A clumsy movement on my part makes it fall off at the moment of extraction. It is all over: the thing will not hatch; I shall not be able to observe the development of the larva. The Tarantula lies motionless, flexible as in life, with not a trace of a wound. In short, we have here life without movement. From time to time the tips of the tarsi quiver a little; and that is all. Accustomed of old to these deceptive corpses, I can see in my mind's eye what has happened: the Spider has been stung in the region of the thorax, no doubt once only, in view of the concentration of her nervous system. I place the victim in a box in which it retains all the pliancy and all the freshness of life from the 2nd of August to the 20th of September, that is to say, for seven weeks. These miracles are familiar to us; there is no need to linger over them here.

The most important matter has escaped me. What I wanted, what I still want to see is the Pompilus engaged in mortal combat with the Lycosa. What a duel, in which the cunning of the one has to overcome the terrible weapons of the other! Does the Wasp enter the burrow to surprise the Tarantula at the bottom of her lair? Such temerity would be fatal to her. Where the big bumblebee dies an instant death, the audacious visitor would perish the moment she entered. Is not the other there, facing her, ready to snap at the back of her head, inflicting a wound which would result in sudden death? No, the Pompilus does not enter the Spider's parlour, that is obvious. Does she surprise the Spider outside her fortress? But the Lycosa is a stay-at-home animal; I do not see her straying abroad during the summer. Later, in the autumn, when the Pompili have disappeared, She wanders about; turning gipsy, she takes the open air with her numerous family, which she carries on her back. Apart from these maternal strolls, she does not appear to me to leave her castle; and the Pompilus, I should think, has no great chance of meeting her outside. The problem, we perceive, is becoming complicated: the huntress cannot make her way into the burrow, where she would risk sudden death; and the Spider's sedentary habits make an encounter outside the burrow improbable. Here is a riddle which would be interesting to decipher. Let us endeavour to do so by observing other Spider-hunters; analogy will enable us to draw a conclusion.

I have often watched Pompili of every species on their hunting-expeditions, but I have never surprised them entering the Spider's lodging when the latter was at home. Whether this lodging be a funnel plunging its neck into a hole in some wall, an awning stretched amid the stubble, a tent modelled upon the Arab's, a sheath formed of a few leaves bound together, or a net with a guardroom attached, whenever the owner is indoors the suspicious Pompilus holds aloof. When the dwelling is vacant, it is another matter: the Wasp moves with arrogant ease over those webs, springes and cables in which so many other insects would remain ensnared. The silken threads do not seem to have any hold upon her. What is she doing, exploring those empty webs? She is watching to see what is happening on the adjacent webs where the Spider is ambushed. The Pompilus therefore feels an insuperable reluctance to make straight for the Spider when the latter is at home in the midst of her snares. And she is right, a hundred times over. If the Tarantula understands the practice of the dagger-thrust in the neck, which is immediately fatal, the other cannot be unacquainted with it. Woe then to the imprudent Wasp who presents herself upon the threshold of a Spider of approximately equal strength!

Of the various instances which I have collected of this cautious reserve on the Spider-huntress' part I will confine myself to the following, which will be sufficient to prove my point. By joining, with silken strands, the three folioles which form the leaf of Virgil's cytisus, a Spider has built herself a green arbour, a horizontal sheath, open at either end. A questing Pompilus comes upon the scene, finds the game to her liking and pops in her head at the entrance of the cell. The Spider immediately retreats to the other end. The huntress goes round the Spider's dwelling and reappears at the other door. Again the Spider retreats, returning to the first entrance. The Wasp also returns to it, but always by the outside. Scarcely has she done so, when the Spider rushes for the opposite opening; and so on for fully a quarter of an hour, both of them coming and going from one end of the cylinder to the other, the Spider inside and the Pompilus outside.

The quarry was a valuable one, it seems, since the Wasp persisted for a long time in her attempts, which were invariably defeated; however, the huntress had to abandon them, baffled by this perpetual running to and fro. The Pompilus made off; and the Spider, once more on the watch, patiently awaited the heedless Midges. What should the Wasp have done to capture this much-coveted game? She should have entered the verdant cylinder, the Spider's dwelling, and pursued the Spider direct, in her own house, instead of remaining outside, going from one door to the other. With such swiftness and dexterity as hers, it seemed to me impossible that the stroke should fail: the quarry moved clumsily, a little sideways, like a Crab. I judged it to be an easy matter; the Pompilus thought it highly dangerous. today I am of her opinion: if she had entered the leafy tube, the mistress of the house would have operated on her neck and the huntress would have become the quarry.

Years passed and the paralyser of the Spiders still refused to reveal her secret; I was badly served by circumstances, could find no leisure, was absorbed in unrelenting preoccupations. At length, during my last year at Orange, the light dawned upon me. My garden was enclosed by an old wall, blackened and ruined by time, where, in the chinks between the stones, lived a population of Spiders, represented more particularly by Segestria perfidia. This is the common Black Spider, or Cellar Spider. She is deep black all over, excepting the mandibles, which are a splendid metallic green. Her two poisoned daggers look like a product of the metalworker's art, like the finest bronze. In any mass of abandoned masonry there is not a quiet corner, not a hole the size of one's finger, in which the Segestria does not set up house. Her web is a widely flaring funnel, whose open end, at most a span across, lies spread upon the surface of the wall, where it is held in place by radiating threads. This conical surface is continued by a tube which runs into a hole in the wall. At the end is the dining-room to which the Spider retires to devour at her ease her captured prey.

With her two hind-legs stuck into the tube to obtain a purchase and the six others spread around the orifice, the better to perceive on every side the quiver which gives the signal of a capture, the Segestria waits motionless, at the entrance of her funnel, for an insect to become entangled in the snare. Large Flies, Drone-flies, dizzily grazing some thread of the snare with their wings, are her usual victims. At the first flutter of the netted Fly, the Spider runs or even leaps forward, but she is now secured by a cord which escapes from the spinnerets and which has its end fastened to the silken tube. This prevents her from falling as she darts along a vertical surface. Bitten at the back of the head, the Drone-fly is dead in a moment; and the Segestria carries him into her lair.

Thanks to this method and these hunting-appliances – an ambush at the bottom of a silken whirlpool, radiating snares, a lifeline which holds her from behind and allows her to take a sudden rush without risking a fall – the Segestria is able to catch game less inoffensive than the Drone-fly. A Common Wasp, they tell me, does not daunt her. Though I have not tested this, I readily believe it, for I well know the Spider's boldness.

This boldness is reinforced by the activity of the venom. It is enough to have seen the Segestria capture some large Fly to be convinced of the overwhelming effect of her fangs upon the insects bitten in the neck. The death of the Drone-fly, entangled in the silken funnel, is reproduced by the sudden death of the bumblebee on entering the Tarantula's burrow. We know the effect of the poison on man, thanks to Antoine Duges' investigations. Let us listen to the brave experimenter:

"The treacherous Segestria, or Great Cellar Spider, reputed poisonous in our part of the country, was chosen for the principal subject of our experiments. She was three-quarters of an inch long, measured from the mandibles to the spinnerets. Taking her in my fingers from behind, by the legs, which were folded and gathered together (this is the way to catch hold of live Spiders, if you would avoid their bite and master them without mutilating them), I placed her on various objects and on my clothes, without her manifesting the least desire to do any harm; but hardly was she laid on the bare skin of my forearm when she seized a fold of the epidermis in her powerful mandibles, which are of a metallic green, and drove her fangs deep into it. For a few moments she remained hanging, although left free; then she released herself, fell and fled, leaving two tiny wounds, a sixth of an inch apart, red, but hardly bleeding, with a slight extravasation round the edge and resembling the wounds produced by a large pin.

"At the moment of the bite, the sensation was sharp enough to deserve the name of pain; and this continued for five or six minutes more, but not so forcibly. I might compare it with the sensation produced by the stinging-nettle. A whitish tumefaction almost immediately surrounded the two pricks; and the circumference, within a radius of about an inch, was coloured an erysipelas red, accompanied by a very slight swelling. In an hour and a half, it had all disappeared, except the mark of the pricks, which persisted for several days, as any other small wound would have done. This was in September, in rather cool weather. Perhaps the symptoms would have displayed somewhat greater severity at a warmer season."

Without being serious, the effect of the Segestria's poison is plainly marked. A sting causing sharp pain and swelling, with the redness of erysipelas, is no trifling matter. While Duges' experiment reassures us in so far as we ourselves are concerned, it is none the less the fact that the Cellar Spider's poison is a terrible thing for insects, whether because of the small size of the victim, or because it acts with special efficacy upon an organization which differs widely from our own. One Pompilus, though greatly inferior to the Segestria in size and strength, nevertheless makes war upon the Black Spider and succeeds in overpowering this formidable quarry. This is Pompilus apicalis, VAN DER LIND, who is hardly larger than the Hive-bee, but very much slenderer. She is of a uniform black; her wings are a cloudy brown, with transparent tips. Let us follow her in her expeditions to the old wall inhabited by the Segestria: we will track her for whole afternoons during the July heats; and we will arm ourselves with patience, for the perilous capture of the game must take the Wasp a long time.

The Spider-huntress explores the wall minutely; she runs, leaps and flies; she comes and goes, flitting to and fro. The antennae quiver; the wings, raised above the back, continually beat one against the other. Ah, here she is, close to a Segestria's funnel! The Spider, who has hitherto remained invisible, instantly appears at the entrance to the tube; she spreads her six forelegs outside, ready to receive the huntress. Far from fleeing before the terrible apparition, she watches the watcher, fully prepared to prey upon her enemy. Before this intrepid demeanour the Pompilus draws back. She examines the coveted game, walks round it for a moment, then goes away without attempting anything. When she has gone, the Segestria retires indoors, backwards. For the second time the Wasp passes near an inhabited funnel. The Spider on the lookout at once shows herself on the threshold of her dwelling, half out of her tube, ready for defence and perhaps also for attack. The Pompilus moves away and the Segestria reenters her tube. A fresh alarm: the Pompilus returns; another threatening demonstration on the part of the Spider. Her neighbour, a little later, does better than this: while the huntress is prowling about in the neighbourhood of the funnel, she suddenly leaps out of the tube, with the lifeline which will save her from falling, should she miss her footing, attached to her spinnerets; she rushes forward and hurls herself in front of the Pompilus, at a distance of some eight inches from her burrow. The Wasp, as though terrified, immediately decamps; and the Segestria no less suddenly retreats indoors.

Here, we must admit, is a strange quarry: it does not hide, but is eager to show itself; it does not run away, but flings itself in front of the hunter. If our observations were to cease here, could we say which of the two is the hunter and which the hunted? Should we not feel sorry for the imprudent Pompilus? Let a thread of the trap entangle her leg; and it is all up with her. The other will be there, stabbing her in the throat. What then is the method which she employs against the Segestria, always on the alert, ready for defence, audacious to the point of aggression? Shall I surprise the reader if I tell him that this problem filled me with the most eager interest, that it held me for weeks in contemplation before that cheerless wall? Nevertheless, my tale will be a short one.

On several occasions I see the Pompilus suddenly fling herself on one of the Spider's legs, seize it with her mandibles and endeavour to draw the animal from its tube. It is a sudden rush, a surprise attack, too quick to permit the Spider to parry it. Fortunately, the latter's two hind-legs are firmly hooked to the dwelling; and the Segestria escapes with a jerk, for the other, having delivered her shock attack, hastens to release her hold; if she persisted, the affair might end badly for her. Having failed in this assault, the Wasp repeats the procedure at other funnels; she will even return to the first when the alarm is somewhat assuaged. Still hopping and fluttering, she prowls around the mouth, whence the Segestria watches her, with her legs outspread. She waits for the propitious moment; she leaps forward, seizes a leg, tugs at it and springs out of reach. More often than not, the Spider holds fast; sometimes she is dragged out of the tube, to a distance of a few inches, but immediately returns, no doubt with the aid of her unbroken lifeline.

The Pompilus' intention is plain: she wants to eject the Spider from her fortress and fling her some distance away. So much perseverance leads to success. This time all goes well: with a vigorous and well-timed tug the Wasp has pulled the Segestria out and at once lets her drop to the ground. Bewildered by her fall and even more demoralized by being wrested from her ambush, the Spider is no longer the bold adversary that she was. She draws her legs together and cowers into a depression in the soil. The huntress is there on the instant to operate on the evicted animal. I have barely time to draw near to watch the tragedy when the victim is paralysed by a thrust of the sting in the thorax.

Here at last, in all its Machiavellian cunning, is the shrewd method of the Pompilus. She would be risking her life if she attacked the Segestria in her home; the Wasp is so convinced of it that she takes good care not to commit this imprudence; but she knows also that, once dislodged from her dwelling, the Spider is as timid, as cowardly as she was bold at the centre of her funnel. The whole point of her tactics, therefore, lies in dislodging the creature. This done, the rest is nothing.

The Tarantula-huntress must behave in the same manner. Enlightened by her kinswoman, Pompilus apicalis, my mind pictures her wandering stealthily around the Lycosa's rampart. The Lycosa hurries up from the bottom of her burrow, believing that a victim is approaching; she ascends her vertical tube, spreading her forelegs outside, ready to leap. But it is the Ringed Pompilus who leaps, seizes a leg, tugs and hurls the Lycosa from her burrow. The Spider is henceforth a craven victim, who will let herself be stabbed without dreaming of employing her venomous fangs. Here craft triumphs over strength; and this craft is not inferior to mine, when, wishing to capture the Tarantula, I make her bite a spike of grass which I dip into the burrow, lead her gently to the surface and then with a sudden jerk throw her outside. For the entomologist as for the Pompilus, the essential thing is to make the Spider leave her stronghold. After this there is no difficulty in catching her, thanks to the utter bewilderment of the evicted animal.

Two contrasting points impress me in the facts which I have just set forth: the shrewdness of the Pompilus and the folly of the Spider. I will admit that the Wasp may gradually have acquired, as being highly beneficial to her posterity, the instinct by which she first of all so judiciously drags the victim from its refuge, in order there to paralyse it without incurring danger, provided that you will explain why the Segestria, possessing an intellect no less gifted than that of the Pompilus, does not yet know how to counteract the trick of which she has so long been the victim. What would the Black Spider need to do to escape her exterminator? Practically nothing: it would be enough for her to withdraw into her tube, instead of coming up to post herself at the entrance, like a sentry, whenever the enemy is in the neighbourhood. It is very brave of her, I agree, but also very risky. The Pompilus will pounce upon one of the legs spread outside the burrow for defence and attack; and the besieged Spider will perish, betrayed by her own boldness. This posture is excellent when waiting for prey.

But the Wasp is not a quarry; she is an enemy and one of the most dreaded of enemies. The Spider knows this. At the sight of the Wasp, instead of placing herself fearlessly but foolishly on her threshold, why does she not retreat into her fortress, where the other would not attack her? The accumulated experience of generations should have taught her this elementary tactical device, which is of the greatest value to the prosperity of her race. If the Pompilus has perfected her method of attack, why has not the Segestria perfected her method of defence? Is it possible that centuries upon centuries should have modified the one to its advantage without succeeding in modifying the other? Here I am utterly at a loss. And I say to myself, in all simplicity: since the Pompili must have Spiders, the former have possessed their patient cunning and the other their foolish audacity from all time. This may be puerile, if you like to think it so, and not in keeping with the transcendental aims of our fashionable theorists; the argument contains neither the subjective nor the objective point of view, neither adaptation nor differentiation, neither atavism nor evolutionism. Very well, but at least I understand it.

Let us return to the habits of Pompilus apicalis. Without expecting results of any particular interest, for in captivity the respective talents of the huntress and the quarry seem to slumber, I place together, in a wide jar, a Wasp and a Segestria. The Spider and her enemy mutually avoid each other, both being equally timid. A judicious shake or two brings them into contact. The Segestria, from time to time, catches hold of the Pompilus, who gathers herself up as best she can, without attempting to use her sting; the Spider rolls the insect between her legs and even between her mandibles, but appears to dislike doing it. Once I see her lie on her back and hold the Pompilus above her, as far away as possible, while turning her over in her forelegs and munching at her with her mandibles. The Wasp, whether by her own adroitness or owing to the Spider's dread of her, promptly escapes from the terrible fangs, moves to a short distance and does not seem to trouble unduly about the buffeting which she has received. She quietly polishes her wings and curls her antennae by pulling them while standing on them with her fore-tarsi. The attack of the Segestria, stimulated by my shakes, is repeated ten times over; and the Pompilus always escapes from the venomous fangs unscathed, as though she were invulnerable.

Is she really invulnerable? By no means, as we shall soon have proved to us; if she retires safe and sound, it is because the Spider does not use her fangs. What we see is a sort of truce, a tacit convention forbidding deadly strokes, or rather the demoralization due to captivity; and the two adversaries are no longer in a sufficiently warlike mood to make play with their daggers. The tranquillity of the Pompilus, who keeps on jauntily curling her antennae in face of the Segestria, reassures me as to my prisoner's fate; for greater security, however, I throw her a scrap of paper, in the folds of which she will find a refuge during the night. She instals herself there, out of the Spider's reach. Next morning I find her dead. During the night the Segestria, whose habits are nocturnal, has recovered her daring and stabbed her enemy. I had my suspicions that the parts played might be reversed! The butcher of yesterday is the victim of today.

I replace the Pompilus by a Hive-bee. The interview is not protracted. Two hours later, the Bee is dead, bitten by the Spider. A Drone-fly suffers the same fate. The Segestria, however, does not touch either of the two corpses, any more than she touched the corpse of the Pompilus. In these murders the captive seems to have no other object than to rid herself of a turbulent neighbour. When appetite awakes, perhaps the victims will be turned to account. They were not; and the fault was mine. I placed in the jar a bumblebee of average size. A day later the Spider was dead; the rude sharer of her captivity had done the deed.

Let us say no more of these unequal duels in the glass prison and complete the story of the Pompilus whom we left at the foot of the wall with the paralysed Segestria. She abandons her prey on the ground and returns to the wall. She visits the Spider's funnels one by one, walking on them as freely as on the stones; she inspects the silken tubes, dipping her antennae into them, sounding and exploring them; she enters without the least hesitation. Whence does she now derive the temerity thus to enter the Segestria's haunts? But a little while ago, she was displaying extreme caution; at this moment, she seems heedless of danger. The fact is that there is no danger really. The Wasp is inspecting uninhabited houses. When she dives down a silken tunnel, she very well knows that there is no one in, for, had the Segestria been there, she would by this time have appeared on the threshold. The fact that the householder does not show herself at the first vibration of the neighbouring threads is a certain proof that the tube is vacant; and the Pompilus enters in full security. I would recommend future observers not to take the present investigations for hunting-tactics. I have already remarked and I repeat: the Pompilus never enters the silken ambush while the Spider is there.

Among the funnels inspected one appears to suit her better than the others; she returns to it frequently in the course of her investigations, which last for nearly an hour. From time to time she hastens back to the Spider lying on the ground; she examines her, tugs at her, drags her a little closer to the wall, then leaves her the better to reconnoitre the tunnel which is the object of her preference. Lastly she returns to the Segestria and takes her by the tip of the abdomen. The quarry is so heavy that she has great difficulty in moving it along the level ground. Two inches divide it from the wall. She gets to the wall, not without effort; nevertheless, once the wall is reached, the job is quickly done. We learn that Antaeus, the son of Mother Earth, in his struggle with Hercules, received new strength as often as his feet touched the ground; the Pompilus, the daughter of the wall, seems to increase her powers tenfold once she has set foot on the masonry.

For here is the Wasp hoisting her prey backwards, her enormous prey, which dangles beneath her. She climbs now a vertical plane, now a slope, according to the uneven surface of the stones. She crosses gaps where she has to go belly uppermost, while the quarry swings to and fro in the air. Nothing stops her; she keeps on climbing, to a height of six feet or more, without selecting her path, without seeing her goal, since she goes backwards. A lodge appears no doubt reconnoitred beforehand and reached, despite the difficulties of an ascent which did not allow her to see it. The Pompilus lays her prey on it. The silken tube which she inspected so lovingly is only some eight inches distant. She goes to it, examines it rapidly and returns to the Spider, whom she at length lowers down the tube.

Shortly afterwards I see her come out again. She searches here and there on the wall for a few scraps of mortar, two or three fairly large pieces, which she carries to the tube, to close it up. The task is done. She flies away.

Next day I inspect this strange burrow. The Spider is at the bottom of the silken tube, isolated on every side, as though in a hammock. The Wasp's egg is glued not to the ventral surface of the victim but to the back, about the middle, near the beginning of the abdomen. It is white, cylindrical and about a twelfth of an inch long. The few bits of mortar which I saw carried have but very roughly blocked the silken chamber at the end. Thus Pompilus apicalis lays her quarry and her eggs not in a burrow of her own making, but in the Spider's actual house. Perhaps the silken tube belongs to this very victim, which in that event provides both board and lodging. What a shelter for the larva of this Pompilus: the warm retreat and downy hammock of the Segestria!

Here then, already, we have two Spider-huntresses, the Ringed Pompilus and P. apicalis, who, unversed in the miner's craft, establish their offspring inexpensively in accidental chinks in the walls, or even in the lair of the Spider on whom the larva feeds. In these cells, acquired without exertion, they build only an attempt at a wall with a few fragments of mortar. But we must beware of generalizing about this expeditious method of establishment. Other Pompili are true diggers, valiantly sinking a burrow in the soil, to a depth of a couple of inches. These include the Eight-spotted Pompilus (P. octopunctatus, PANZ.), with her black-and-yellow livery and her amber wings, a little darker at the tips. For her game she chooses the Epeirae, those fat Spiders, magnificently adorned, who lie in wait at the centre of their large, vertical webs. I am not sufficiently acquainted with her habits to describe them; above all, I know nothing of her hunting-tactics. But her dwelling is familiar to me: it is a burrow, which I have seen her begin, complete and close according to the customary method of the Digger-wasps.

CHAPTER 2.

THE SCOLIAE.

***

Were strength to take precedence over the other zoological attributes, the Scoliae would hold a predominant place in the front rank of the Wasps. Some of them may be compared in size with the little bird from the north, the Golden-crested Wren, who comes to us at the time of the first autumn mists and visits the rotten buds. The largest and most imposing of our sting-bearers, the Carpenter-bee, the bumblebee, the Hornet, cut a poor figure beside certain of the Scoliae. Of this group of giants my district possesses the Garden Scolia (S. hortorum, VAN DER LIND), who is over an inch and a half in length and measures four inches from tip to tip of her outspread wings, and the Hemorrhoidal Scolia (S. haemorrhoidalis, VAN DER LIND), who rivals the Garden Scolia in point of size and is distinguished more particularly by the bundle of red hairs bristling at the tip of the abdomen.

A black livery, with broad yellow patches; leathery wings, amber-coloured, like the skin of an onion, and watered with purple reflections; thick, knotted legs, covered with sharp hairs; a massive frame; a powerful head, encased in a hard cranium; a stiff, clumsy gait; a low, short, silent flight: this gives you a concise description of the female, who is strongly equipped for her arduous task. The male, being a mere philanderer, sports a more elegant pair of horns, is more daintily clad and has a more graceful figure, without altogether losing the quality of robustness which is his consort's leading characteristic.

It is not without a certain alarm that the insect-collector finds himself for the first time confronted by the Garden Scolia. How is he to capture the imposing creature, how to avoid its sting? If its effect is in proportion to the Wasp's size, the sting of the Scolia must be something terrible. The Hornet, though she unsheath her weapon but once, causes the most exquisite pain. What would it be like if one were stabbed by this colossus? The prospect of a swelling as big as a man's fist and as painful as the touch of a red-hot iron passes through our mind at the moment when we are bringing down the net. And we refrain, we beat a retreat, we are greatly relieved not to have aroused the dangerous creature's attention.

Yes, I confess to having run away from my first Scoliae, anxious though I was to enrich my budding collection with this magnificent insect. There were painful recollections of the Common Wasp and the Hornet connected with this excess of prudence. I say excess, for today, instructed by long experience, I have quite recovered from my former fears; and, when I see a Scolia resting on a thistle-head, I do not scruple to take her in my fingers, without any precaution whatever, however large she may be and however menacing her aspect. My courage is not all that it seems to be; I am quite ready to tell the Wasp-hunting novice this. The Scoliae are notably peaceable. Their sting is an implement of labour far more than a weapon of war; they use it to paralyse the prey destined for their offspring; and only in the last extremity do they employ it in self-defence. Moreover, the lack of agility in their movements nearly always enables us to avoid their sting; and, even if we be stung, the pain is almost insignificant. This absence of any acute smarting as a result of the poison is almost constant in the Hunting Wasps, whose weapon is a surgical lancet and devised for the most delicate physiological operations.

Among the other Scoliae of my district I will mention the Two-banded Scolia (S. bifasciata, VAN DER LIND), whom I see every year, in September, working at the heaps of leaf-mould which are placed for her benefit in a corner of my paddock; and the Interrupted Scolia (S. interrupta, LATR.), the inhabitant of the sandy soil at the foot of the neighbouring hills. Much smaller than the two preceding insects, but also much commoner, a necessary condition of continuous observation, they will provide me with the principal data for this study of the Scoliae.

I open my old note book; and I see myself once more, on the 6th of August, 1857, in the Bois des Issards, that famous copse near Avignon which I have celebrated in my essay on the Bembex-wasps. Once again, my head crammed with entomological projects, I am at the beginning of my holidays which, for two months, will allow me to indulge in the insect's company.

A fig for Mariotte's flask and Toricelli's tube! This is the thrice-blest period when I cease to be a schoolmaster and become a schoolboy, the schoolboy in love with animals. Like a madder-cutter off for his day's work, I set out carrying over my shoulder a solid digging-implement, the local luchet, and on my back my game-bag with boxes, bottles, trowel, glass tubes, tweezers, lenses and other impedimenta. A large umbrella saves me from sunstroke. It is the most scorching hour of the hottest day in the year. Exhausted by the heat, the Cicadae are silent. The bronze-eyed gadflies seek a refuge from the pitiless sun under the roof of my silken shelter; other large Flies, the sobre-hued Pangoniae, dash themselves recklessly against my face.

The spot at which I have installed myself is a sandy clearing which I had recognized the year before as a site beloved of the Scoliae. Here and there are scattered thickets of holm-oak, whose dense undergrowth shelters a bed of dead leaves and a thin layer of mould. My memory has served me well. Here, sure enough, as the heat grows a little less, appear, coming I know not from whence, some Two-banded Scoliae. The number increases; and it is not long before I see very nearly a dozen of them about me, close enough for observation. By their smaller size and more buoyant flight, they are easily known for males. Almost grazing the ground, they fly softly, going to and fro, passing and repassing in every direction. From time to time one of them alights on the ground, feels the sand with his antennae and seems to be enquiring into what is happening in the depths of the soil; then he resumes his flight, alternately coming and going.

What are they waiting for? What are they seeking in these evolutions of theirs, which are repeated a hundred times over? Food? No, for close beside them stand several eryngo-stems, whose sturdy clusters are the Wasps' usual resource at this season of parched vegetation; and not one of them settles upon the flowers, not one of them seems to care about their sugary exudations. Their attention is engrossed elsewhere. It is the ground, it is the stretch of sand which they are so assiduously exploring; what they are waiting for is the arrival of some female, who bursting the cocoon, may appear from one moment to the next, issuing all dusty from the ground. She will not be given time to brush herself or to wash her eyes: three or four more of them will be there at once, eager to dispute her possession. I am too familiar with the amorous contests of the Hymenopteron clan to allow myself to be mistaken. It is the rule for the males, who are the earlier of the two, to keep a close guard around the natal spot and watch for the emergence of the females, whom they pester with their pursuit the moment they reach the light of day. This is the motive of the interminable ballet of my Scoliae. Let us have patience: perhaps we shall witness the nuptials.

The hours go by; the Pangoniae and the gadflies desert my umbrella; the Scoliae grow weary and gradually disappear. It is finished. I shall see nothing more today. I repeat my laborious expedition to the Bois des Issards over and over again; and each time I see the males as assiduous as ever in skimming over the ground. My perseverance deserved to succeed. It did, though the success was very incomplete. Let me describe it, such as it was; the future will fill up the gaps.

A female issues from the soil before my eyes. She flies away, followed by several males. With the luchet I dig at the point of emergence; and, as the excavation progresses, I sift between my fingers the rubbish of sand mixed with mould. In the sweat of my brow, as I may justly say, I must have removed nearly a cubic yard of material, when at last I make a find. This is a recently ruptured cocoon, to the side of which adheres an empty skin, the last remnant of the game on which the larva fed that wrought the said cocoon. Considering the good condition of its silken fabric, this cocoon may have belonged to the Scolia who has just quitted her underground dwelling before my eyes. As for the skin accompanying it, this has been so much spoilt by the moisture of the soil and by the grassy roots that I cannot determine its origin exactly. The cranium, however, which is better-preserved, the mandibles and certain details of the general configuration lead me to suspect the larva of a Lamellicorn.

It is getting late. This is enough for today. I am worn out, but amply repaid for my exertions by a broken cocoon and the puzzling skin of a wretched grub. Young people who make a hobby of natural history, would you like to discover whether the sacred fire flows in your veins? Imagine yourselves returning from such an expedition. You are carrying on your shoulder the peasant's heavy spade; your loins are stiff with the laborious digging which you have just finished in a crouching position; the heat of an August afternoon has set your brain simmering; your eyelids are tired by the itch of an inflammation resulting from the overpowering light in which you have been working; you have a devouring thirst; and before you lies the dusty prospect of the miles that divide you from your well-earned rest. Yet something stings within you; forgetful of your present woes you are absolutely glad of your excursion. Why? Because you have in your possession a shred of rotten skin. If this is so, my young friends, you may go ahead, for you will do something, though I warn you that this does not mean, by a long way, that you will get on in the world.

I examined this shred of skin with all the care that it deserved. My first suspicions were confirmed: a Lamellicorn, a Scarabaeid in the larval state, is the first food of the Wasp whose cocoon I have just unearthed.

But which of the Scarabaeidae? And does this cocoon, my precious booty, really belong to the Scoliae? The problem is beginning to take shape. To attempt its solution we must go back to the Bois des Issards.

I did go back and so often that my patience ended by being exhausted before the problem of the Scoliae had received a satisfactory solution. The difficulties are great indeed, under the conditions. Where am I to dig in the indefinite stretch of sandy soil to light upon a spot frequented by the Scoliae? The luchet is driven into the ground at random; and almost invariably I find none of what I am seeking. To be sure, the males, flying level with the ground, give me a hint, at the outset, with their certainty of instinct, as to the spots where the females ought to be; but their hints are very vague, because they go so far in every direction. If I wished to examine the soil which a single male explores in his flight, with its constantly changing course, I should have to turn over, to the depth of perhaps a yard, at least four poles of earth. This is too much for my strength and the time at my disposal. Then, as the season advances, the males disappear, whereupon I am suddenly deprived of their hints. To know more or less where I should thrust my luchet, I have only one resource left, which is to watch for the females emerging from the ground or else entering it. With a great expenditure of time and patience I have at last had this windfall, very rarely, I admit.

The Scoliae do not dig a burrow which can be compared with that of the other Hunting Wasps; they have no fixed residence, with an unimpeded gallery opening on the outer world and giving access to the cells, the abodes of the larvae. They have no entrance - and exit-doors, no corridor built in advance. If they have to make their way underground, any point not hitherto turned over serves their purpose, provided that it be not too hard for their digging-tools, which, for that matter, are very powerful; if they have to come out, the point of exit is no less indifferent. The Scolia does not bore the soil through which she passes: she excavates and ploughs it with her legs and forehead; and the stuff shifted remains where it lies, behind her, forthwith blocking the passage which she has followed. When she is about to emerge into the outer world, her advent is heralded by the fresh soil which heaps itself into a mound as though heaved up by the snout of some tiny Mole. The insect sallies forth; and the mound collapses, completely filling up the exit-hole. If the Wasp is entering the ground, the digging-operations, undertaken at an arbitrary point, quickly yield a cavity in which the Scolia disappears, separated from the surface by the whole track of shifted material.

I can easily trace her passage through the thickness of the soil by certain long, winding cylinders, formed of loose materials in the midst of compact and stable earth. These cylinders are numerous; they sometimes run to a depth of twenty inches; they extend in all directions, fairly often crossing one another. Not one of them ever exhibits so much as a suspicion of an open gallery. They are obviously not permanent ways of communication with the outer world, but hunting-trails which the insect has followed once, without going back to them. What was the Wasp seeking when she riddled the soil with these tunnels which are now full of running sands? No doubt the food for her family, the larva of which I possess the empty skin, now an unrecognizable shred.

I begin to see a little light: the Scoliae are underground workers. I already expected as much, having before now captured Scoliae soiled with little earthy encrustations on the joints of the legs. The Wasp, who is so careful to keep clean, taking advantage of the least leisure to brush and polish herself, could never display such blemishes unless she were a devoted earth-worker. I used to suspect their trade, now I know it. They live underground, where they burrow in search of Lamellicorn-grubs, just as the Mole burrows in search of the White Worm. It is even possible that, after receiving the embraces of the males, they but very rarely return to the surface, absorbed as they are by their maternal duties; and this, no doubt, is why my patience becomes exhausted in watching for their entrance and their emergence.

It is in the subsoil that they establish themselves and travel to and fro; with the help of their powerful mandibles, their hard cranium, their strong, prickly legs, they easily make themselves paths in the loose earth. They are living ploughshares. By the end of August, therefore, the female population is for the most part underground, busily occupied in egg-laying and provisioning. Everything seems to tell me that I should watch in vain for the appearance of a few females in the broad daylight; I must resign myself to excavating at random.

The result was hardly commensurate with the labour which I expended on digging. I found a few cocoons, nearly all broken, like the one which I already possessed, and, like it, bearing on their side the tattered skin of a larva of the same Scarabaeid. Two of these cocoons which are still intact contained a dead adult

Wasp. This was actually the Two-banded Scolia, a precious discovery which changed my suspicions into a certainty.

I also unearthed some cocoons, slightly different in appearance, containing an adult inmate, likewise dead, in whom I recognized the Interrupted Scolia. The remnants of the provisions again consisted of the empty skin of a larva, also a Lamellicorn, but not the same as the one hunted by the first Scolia. And this was all. Now here, now there, I shifted a few cubic yards of soil, without managing to find fresh provisions with the egg or the young larva. And yet it was the right season, the egg-laying season, for the males, numerous at the outset, had grown rarer day by day until they disappeared entirely. My lack of success was due to the uncertainty of my excavations, in which I had nothing to guide me over the indefinite area covered.

If I could at least identify the Scarabaeidae whose larvae form the prey of the two Scoliae, the problem would be half solved. Let us try. I collect all that the luchet has turned up: larvae, nymphs and adult Beetles. My booty comprises two species of Lamellicorns: Anoxia villosa and Euchlora Julii, both of whom I find in the perfect state, usually dead, but sometimes alive. I obtain a few of their nymphs, a great piece of luck, for the larval skin which accompanies them will serve me as a standard of comparison. I come upon plenty of larvae, of all ages. When I compare them with the cast garment abandoned by the nymphs, I recognize some as belonging to the Anoxia and the rest to the Euchlora.

With these data, I perceive with absolute certainty that the empty skin adhering to the cocoon of the Interrupted

Scolia belongs to the Anoxia. As for the Euchlora, she is not involved in the problem: the larva hunted by the Two-banded Scolia does not belong to her any more than it belongs to the Anoxia. Then with which Scarabaeid does the empty skin which is still unknown to me correspond? The Lamellicorn whom I am seeking must exist in the ground which I have been exploring, because the Two-banded Scolia has established herself there. Later - oh, very long afterwards! - I recognized where my search was at fault. In order not to find a network of roots beneath my luchet and to render the work of excavation lighter, I was digging the bare places, at some distance from the thickets of holm-oak; and it was just in those thickets, which are rich in vegetable mould, that I should have sought. There, near the old stumps, in the soil consisting of dead leaves and rotting wood, I should certainly have come upon the larva so greatly desired, as will be proved by what I have still to say.

Here ends what my earlier investigations taught me. There is reason to believe that the Bois des Issards would never have furnished me with the precise data, in the form in which I wanted them. The remoteness of the spot, the fatigue of the expeditions, which the heat rendered intensely exhausting, the impossibility of knowing which points to attack would undoubtedly have discouraged me before the problem had advanced a step farther. Studies such as these call for home leisure and application, for residence in a country village. You are then familiar with every spot in your own grounds and the surrounding country and you can go to work with certainty.

Twenty-three years have passed; and here I am at Serignan, where I have become a peasant, working by turns on my writing-pad and my cabbage-patch. On the 14th of August, 1880, Favier clears away a heap of mould consisting of vegetable refuse and of leaves stacked in a corner against the wall of the paddock. This clearance is considered necessary because Bull, when the lovers' moon arrives, uses this hillock to climb to the top of the wall and thence to repair to the canine wedding the news of which is brought to him by the effluvia borne upon the air. His pilgrimage fulfilled, he returns, with a discomfited look and a slit ear, but always ready, once he has had his feed, to repeat the escapade. To put an end to this licentious behaviour, which has cost him so many gaping wounds, we decided to remove the heap of soil which serves him as a ladder of escape.

Favier calls me while in the midst of his labours with the spade and barrow:

"Here's a find, sir, a great find! Come and look."

I hasten to the spot. The find is a magnificent one indeed and of a nature to fill me with delight, awakening all my old recollections of the Bois des Issards. Any number of females of the Two-banded Scolia, disturbed at their work, are emerging here and there from the depth of the soil. The cocoons also are plentiful, each lying next to the skin of the victim on which the larva has fed. They are all open but still fresh: they date from the present generation; the Scoliae whom I unearth have quitted them not long since. I learnt later, in fact, that the hatching took place in the course of July.

In the same heap of mould is a swarming colony of Scarabaeidae in the form of larvae, nymphs and adult insects. It includes the largest of our Beetles, the common Rhinoceros Beetle, or Oryctes nasicornis. I find some who have been recently liberated, whose wing-cases, of a glossy brown, now see the sunlight for the first time; I find others enclosed in their earthen shell, almost as big as a Turkey's egg. More frequent is her powerful larva, with its heavy paunch, bent into a hook. I note the presence of a second bearer of the nasal horn, Oryctes Silenus, who is much smaller than her kinswoman, and of Pentodon punctatus, a Scarabaeid who ravages my lettuces.

But the predominant population consists of Cetoniae, or Rosechafers, most of them enclosed in their egg-shaped shells, with earthen walls encrusted with dung. There are three different species: C. aurata, C. morio and C. floricola. Most of them belong to the first species. Their larvae, which are easily recognized by their singular talent for walking on their backs with their legs in the air, are numbered by the hundred. Every age is represented, from thc ncw born grub to the podgy larva on the point of building its shell.

This time the problem of the victuals is solved. When I compare the larval slough sticking to the Scolia's cocoons with the Cetonia-larvae or, better, with the skin cast by these larvae, under cover of the cocoon, at the moment of the nymphal transformation, I establish an absolute identity. The Two-banded Scolia rations each of her eggs with a Cetonia-grub. Behold the riddle which my irksome searches in the Bois des Issards had not enabled me to solve. today, at my threshold, the difficult problem becomes child's play. I can investigate the question easily to the fullest possible extent; I need not put myself out at all; at any hour of the day, at any period that seems favourable, I have the requisite elements before my eyes. Ah, dear village, so poor, so countrified, how happily inspired was I when I came to ask of you a hermit's retreat, where I could live in the company of my beloved insects and, in so doing, set down not too unworthily a few chapters of their wonderful history!

According to the Italian observer Passerini, the Garden Scolia feeds her family on the larvae of Oryctes nasicornis, in the heaps of old tan-waste removed from the hothouses. I do not despair of seeing this colossal Wasp coming to establish herself one day in my heaps of leaf-mould, in which the same Scarabaeid is swarming. Her rarity in my part of the country is probably the only cause that has hitherto prevented the realization of my wishes.

I have just shown that the Two-banded Scolia feeds in infancy on Cetonia-larvae and particularly on those of C. aurata, C. morio and C. floricola. These three species dwell together in the rubbish-heap just explored; their larvae differ so little that I should have to examine them minutely to distinguish the one from the other; and even then I should not be certain of succeeding. It seems probable that the Scolia does not choose between them, that she uses all three indiscriminately. Perhaps she even assails other larvae, inhabitants, like the foregoing, of heaps of rotting vegetable-matter. I therefore set down the Cetonia genus generally as forming the prey of the Two-banded Scolia.

Lastly, round about Avignon, the Interrupted Scolia used to prey upon the larva of the Shaggy Anoxia (A. villosa). At Serignan, which is surrounded by the same kind of sandy soil, without other vegetation than a few sparse seed-bearing grasses, I find her rationing her young with the Morning Anoxia (A. matutinalis). Oryctes, Cetoniae and Anoxiae in the larval state: here then is the prey of the three Scoliae whose habits we know. The three Beetles are Lamellicorns, Scarabaeidae. We shall have occasion later to consider the reason of this very striking coincidence.

For the moment, the business in hand is to move the heap of leaf-mould to some other place, with the wheelbarrow. This is Favier's work, while I myself collect the disturbed population in glass jars, in order to put them back into the new rubbish-heap with all the consideration which my plans owe to them. The laying-time has not yet set in, for I find no eggs, no young Scolia-larvae. September apparently will be the propitious month. But there are bound to be many injured in the course of this upheaval; some of the Scoliae have flown away who will perhaps have a certain difficulty in finding the new site; I have disarranged everything in the overturned heap. To allow tranquility to be restored and habit to resume its rounds, to give the population time to increase and replace the fugitives and the injured, it would be best, I think, to leave the heap alone this year and not to resume my investigations until the next. After the thorough confusion due to the removal, I should jeopardize success by being too precipitate. Let us wait one year more. I decide accordingly, curb my impatience and resign myself. We will simply confine ourselves to enlarging the heap, when the leaves begin to fall, by accumulating the refuse that strews the paddock, so that we may have a richer field of operations.

In the following August, my visits to the mound of leaf-mould become a daily habit. By two o'clock in the afternoon, when the sun has cleared the adjacent pine-trees and is shining on the heap, numbers of male Scoliae arrive from the neighbouring fields, where they have been slaking their thirst on the eryngo-heads. Incessantly coming and going with an indolent flight, they circle round the heap. If some female rise from the soil, those who have seen her dart forward. A not very turbulent affray decides which of the suitors shall be the possessor; and the couple fly away over the wall. This is a repetition of what I used to see in the Bois des Issards. By the time that August is over. The males have ceased to show themselves. The mothers do not appear either: they are busy underground, establishing their families.

On the 2nd of September, I decide upon a search with my son Emile, who handles the fork and the shovel, while I examine the clods dug up. Victory! A magnificent result, finer than any that my fondest ambition would have dared to contemplate! Here is a vast array of Cetonia-larvae, all flaccid, motionless, lying on their backs, with a Scolia's egg sticking to the centre of their abdomen; here are young Scolia-larvae dipping their heads into the entrails of their victims; here are others farther advanced, munching their last mouthfuls of a prey which is drained dry and reduced to a skin; here are some laying the foundation of their cocoons with a reddish silk, which looks as if it had been dyed in Bullock's blood; here are some whose cocoons are finished. There is plenty of everything, from the egg to the larva whose period of activity is over. I mark the 2nd of September as a red-letter day; it has given me the final key to a riddle which has kept me in suspense for nearly half a century.

I place my spoils religiously in shallow, wide-mouthed glass jars containing a layer of finely sifted mould. In this soft bed, which is identical in character with the natal surroundings, I make some faint impressions with my fingers, so many cavities, each of which receives one of my subjects, one only. A pane of glass covers the mouth of the receptacle. In this way I prevent a too rapid evaporation and keep my nurselings under my eyes without fear of disturbing them. Now that all this is in order, let us proceed to record events.

The Cetonia-larvae which I find with a Scolia's egg upon their ventral surface are distributed in the mould at random, without special cavities, without any sign of some sort of structure. They are smothered in the mould, just as are the larvae which have not been injured by the Wasp. As my excavations in the Bois des Issards told me, the Scolia does not prepare a lodging for her family; she knows nothing of the art of cell-building. Her offspring occupies a fortuitous abodc, on which thc mother expends no architectural pains. Whereas the other Hunting Wasps prepare a dwelling to which the provisions are carried, sometimes from a distance, the Scolia confines herself to digging her bed of leaf-mould until she comes upon a Cetonia-larva. When she finds a quarry, she stabs it on the spot, in order to immobilize it; and, again on the spot, she lays an egg on the ventral surface of the paralysed creature. That is all. The mother goes in quest of another prey without troubling further about the egg which has just been laid. There is no effort of carting or building. At the very spot where the Cetonia-grub is caught and paralysed, the Scolia-larva hatches, grows and weaves its cocoon. The establishment of the family is thus reduced to the simplest possible expression.

CHAPTER 3.

A DANGEROUS DIET.

***

The Scolia's egg is in no way exceptional in shape. It is white, cylindrical, straight and about four millimetres long by one millimetre thick. It is fixed, by its fore-end, upon the median line of the victim's abdomen, well to the rear of the legs, near the beginning of the brown patch formed by the mass of food under the skin.

I watch the hatching. The grub, still wearing upon its hinder parts the delicate pellicle which it has just shed, is fixed to the spot to which the egg itself adhered by its cephalic extremity. A striking spectacle, that of the feeble creature, only this moment hatched, boring, for its first mouthful, into the paunch of its enormous prey, which lies stretched upon its back. The nascent tooth takes a day over the difficult task. Next morning the skin has yielded; and I find the newborn larva with its head plunged into a small, round, bleeding wound.

In size the grub is the same as the egg, whose dimensions I have just given. Now the Cetonia-larva, to meet the Scolia's requirements, averages thirty millimetres in length by nine in thickness, whence follows that its bulk is six or seven hundred times as great as that of the newly-hatched grub of the Scolia. Here certainly is a quarry which, were it active and capable of wriggling and biting, would expose the nurseling to terrible attacks. The danger has been averted by the mother's stiletto; and the fragile grub attacks the monster's paunch with as little hesitation as though it were sucking the breast.

Day by day the young Scolia's head penetrates farther into the Cetonia's belly. To pass through the narrow orifice made in the skin, the forepart of the body contracts and lengthens out, as though drawn through a die-plate. The larva thus assumes a rather strange form. Its hinder half, which is constantly outside the victim's belly, has the shape and fulness usual in the larvae of the Digger-wasps, whereas the front half, which, once it has dived under the skin of the exploited victim, does not come out again until the time arrives for spinning the cocoon, tapers off suddenly into a snakelike neck. This front part is moulded, so to speak, by the narrow entrance-hole made in the skin and henceforth retains its slender formation. As a matter of fact, a similar configuration recurs, in varying degrees, in the larvae of the Digger-wasps whose ration consists of a bulky quarry which takes a long time to consume. These include the Languedocian Sphex, with her Ephippiger, and the Hairy Ammophila, with her Grey Worm. There is none of this sudden constriction, dividing the creature into two disparate halves, when the victuals consist of numerous and comparatively small items. The larva then retains its usual shape, being obliged to pass, at brief intervals, from one joint in its larder to the next.

From the first bite of the mandibles, until the whole head of game is consumed, the Scolia-larva is never seen to withdraw its head and its long neck from inside the creature which it is devouring. I suspect the reason of this persistence in attacking a single point; I even seem to perceive the need for a special art in the manner of eating. The Cetonia-larva is a square meal in itself, one large dish, which has to retain a suitable freshness until the end. The young Scolia, therefore, must attack with discretion, at the unvarying point chosen by the mother on the ventral surface, for the entrance-hole is at the exact point where the egg was fixed. As the nurseling's neck lengthens and dives deeper, the victim's entrails are nibbled gradually and methodically: first, the least essential; next, those whose removal leaves yet a remnant of life; lastly, those whose loss inevitably entails death, followed very soon by putrefaction.

At the first bites we see the victim's blood oozing through the wound. It is a highly-elaborated fluid, easy of digestion, and forms a sort of milk-diet for the new-born grub. The little ogre's teat is the bleeding paunch of the Cetonia-larva. The latter will not die of the wound, at least not for some time. The next thing to be tackled is the fatty substance which wraps the internal organs in its delicate folds. This again is a loss which the Cetonia can suffer without dying then and there. Now comes the turn of the muscular layer which lines the skin; now, that of the essential organs; now, that of the nerve-centres and the trachean network, whereupon the last gleam of light is extinguished and the Cetonia reduced to a mere bag, empty but intact, save for the entrance-hole made in the middle of the belly. From now onwards, these remains may rot if they will: the Scolia, by its methodical fashion of consuming its victuals, has succeeded in keeping them fresh to the very last; and now you may see it, replete, shining with health, withdraw its long neck from the bag of skin and prepare to weave the cocoon in which its development will be completed.

It is possible that I may not be quite accurate as to the precise order in which the organs are consumed, for it is not easy to perceive what happens inside the exploited larva's body. The ruling feature in this scientific method of eating, which proceeds from the parts less to the parts more necessary to preserve a remnant of life, is none the less obvious. If direct observation did not already to some degree confirm it, a mere examination of the half-eaten larva would do so in the most positive fashion.

The Cetonia-larva is at first a plump grub. Drained by the Scolia's tooth, it gradually becomes limp and wrinkled. In a few days' time it resembles a shrivelled bit of bacon-fat and then a bag whose two sides have fallen in. Yet this bit of bacon and this bag have the same characteristic look of fresh meat as had the grub before it was bitten into. Despite the persistent nibbling of the Scolia, life continues, holding at bay the inroads of putrefaction until the mandibles have given their last bites. Does not this remnant of tenacious vitality in itself show that the organs of primary importance are the last to be attacked? Does it not prove that there is a progressive dismemberment passing from the less essential to the indispensable?

Would you like to see what becomes of a Cetonia-larva when the organism is wounded in its vital centres at the very beginning? The experiment is an easy one; and I made a point of trying it. A sewing-needle, first softened and flattened into a blade, then retempered and sharpened, gives me a most delicate scalpel. With this instrument I make a fine incision, through which I remove the mass of nerves whose remarkable structure we shall soon have occasion to study. The thing is done: the wound, which does not look serious, has left the creature a corpse, a real corpse. I lay my victim on a bed of moist earth, in a jar with a glass lid; in fact, I establish it in the same conditions as those of the larvae on which the Scoliae feed. By the next day, without changing shape, it has turned a repulsive brown; presently it dissolves into noisome putrescence. On the same bed of earth, under the same glass cover, in the same moist, warm atmosphere, the larvae three-quarters eaten by the Scoliae retain, on the contrary, the appearance of healthy flesh.

If a single stroke of my dagger, fashioned from the point of a needle, results in immediate death and early putrefaction; if the repeated bites of the Scolia gut the creature's body and reduce it almost to a skin without completely killing it, the striking contrast between these two results must be due to the relative importance of the organs injured. I destroy the nerve-centres and inevitably kill my larva, which is putrid by the following day; the Scolia attacks the reserves of fat, the blood, the muscles and does not kill its victim, which will provide it with wholesome food until the end. But it is clear that, if the Scolia were to set to work as I did, there would be nothing left, after the first few bites, but an actual corpse, discharging fluids which would be fatal to it within twenty-four hours. The mother, it is true, in order to assure the immobility of her prey, has injected the poison of her sting into the nerve-centres. Her operation cannot be compared with mine in any respect. She practises the method of the skilful physiologist who induces anaesthesia; I go to work like the butcher who chops, cuts and disembowels. The sting leaves the nerve-centres intact. Deprived of sensibility by the poison, they have lost the power of provoking muscular contractions; but who can say that, numbed as they are, they no longer serve to maintain a faint vitality? The flame is extinguished, but there is still a glowing speck upon the wick. I, a rough blunderer, do more than blow out the lamp: I throw away the wick and all is over. The grub would do the same if it bit straight into the mass of nerves.

Everything confirms the fact: the Scolia and the other Hunting Wasps whose provisions consist of bulky heads of game are gifted with a special art of eating, an exquisitely delicate art which saves a remnant of life in the prey devoured, until it is all consumed. When the prey is a small one, this precaution is superfluous. Consider, for instance, the Bembex-grubs in the midst of their heap of Flies. The prey seized upon is broached on the back, the belly, the head, the thorax, indifferently. The larva munches a given spot, which it leaves to munch a second, passing to a third and a fourth, at the bidding of its changing whims. It seems to taste and select, by repeated trials, the mouthfuls most to its liking. Thus bitton at several points, covered with wounds, the Fly is soon a shapeless mass which would putrefy very quickly if the meagre dish were not devoured at a single meal. Allow the Scolia-grub the same unlicensed gluttony: it would perish beside its corpulent victim, which should have kept fresh for a fortnight, but which almost from the beginning would be no more than a filthy putrescence.

This art of careful eating does not seem easy to practise: at least, the larva, if ever so little diverted from its usual courses, is no longer able to apply its talent as a capable trencherman. This will be proved by experiment. I must begin by observing that, when I spoke of my larva which turned putrid within twenty-four hours, I adopted an extreme case for the sake of greater clearness. The Scolia, taking its first bite, does not and cannot go to such lengths. Nevertheless it behooves us to enquire whether, in the consumption of the victuals, the initial point of attack is a matter of indifference and whether the rummaging through the entrails of the victim entails a determined order, without which success is uncertain or even impossible. To these delicate questions no one, I think, can reply. Where science is silent, perhaps the grub will speak. We will try.

I move from its position a Scolia-grub which has attained a quarter or a third of its full growth. The long neck plunged into the victim's belly is rather difficult to extract, because of the need of molesting the creature as little as possible. I succeed, by means of a little patience and repeated strokes with the tip of a paintbrush. I now turn the Cetonia-larva over, back uppermost, at the bottom of the little hollow made by pressing my finger in the layer of mould. Lastly, I place the Scolia on its victim's back. Here is my grub under the same conditions as just now, with this difference, that the back and not the belly of its victim is presented to its mandibles.

I watch it for a whole afternoon. It writhes about; it moves its little head now in this direction, now in that, frequently laying it on the Cetonia, but without fixing it anywhere. The day draws to a close; and still it has accomplished nothing. There are restless movements, nothing more. Hunger, I tell myself, will eventually induce it to bite. I am wrong. Next morning I find it more anxious than the day before and still groping about, without resolving to fix its mandibles anywhere. I leave it alone for half a day longer without obtaining any result. Yet twenty-four hours of abstinence must have awakened a good appetite, above all in a creature which, if left undisturbed, would not have ceased eating.

Excessive hunger cannot induce it to nibble at an unlawful spot. Is this due to feebleness of the teeth? By no means: the Cetonia's skin is no tougher on the back than on the belly; moreover, the grub is capable of perforating the skin when it leaves the egg; a fortiori, it must be more capable of doing so now that it has attained a sturdy growth. Thus we see no lack of ability, but an obstinate refusal to nibble at a point which ought to be respected. Who knows? On this side perhaps the grub's dorsal vessel would be wounded, its heart, an organ indispensable to life. The fact remains that my attempts to make the grub tackle its victim from the back have failed. Does this mean that it entertains the least suspicion of the danger which it might incur were it to produce putrefaction by awkwardly carving its victuals from the back? It would bc absurd to give such an idea a moment's consideration. Its refusal is dictated by a preordained decree which it is bound to obey.

My Scolia-grubs would die of starvation if I left them on their victim's back. I therefore restore matters as they were, with the Cetonia-larva belly uppermost and the young Scolia on top. I might utilise the subjects of my previous experiments; but, as I have to take precautions against the disturbance which may have been caused by the test already undergone, I prefer to operate on new patients, a luxury in which the richness of my menagerie allows me to indulge. I move the Scolia from its position, extract its head from the entrails of the Cetonia-larva and leave it to its own resources on its victim's belly. Betraying every symptom of uneasiness, the grub gropes, hesitates, casts about and does not insert its mandibles anywhere, though it is now the ventral surface which it is exploring. It would not display greater hesitation if placed on the back of the larva. I repeat, who knows? On this side it might perhaps injure the nervous plexus, which is even more essential than the dorsal vessel. The inexperienced grub must not drive in its mandibles at random; its future is jeopardized if it gives a single ill-judged bite. If it gnaws at the spot where I myself operated with my needle wrought into a scalpel, its victuals will very soon turn putrid. Once more, then, we witness an absolute refusal to perforate the skin of the victim elsewhere than at the very point where the egg was fixed.

The mother selects this point, which is undoubtedly that most favourable to the future prosperity of the larva, though I am not able clearly to discern the reasons for her choice; she fixes the egg to it; and the place where the opening is to be made is henceforth determined. It is here that the grub must bite: only here, never elsewhere. Its invincible refusal to tackle the Cetonia in any other part, even though it should die of starvation, shews us how rigorous is the rule of conduct with which its instinct is inspired.

As it gropes about, the grub laid on the victim's ventral surface sooner or later rediscovers the gaping wound from which I have removed it. If this takes too long for my patience, I can myself guide its head to the place with the point of a paintbrush. The grub then recognizes the hole of its own making, slips its neck into it and little by little dives into the Cetonia's belly, so that the original state of affairs appears to be exactly restored. And yet its successful rearing is henceforth highly problematical.

It is possible that the larva will prosper, complete its development and spin its cocoon; it is also possible – and the case is not unusual – that the Cetonia-larva will soon turn brown and putrid. We then see the Scolia itself turn brown, distended as it is with putrescent foodstuffs, and then cease all movement, without attempting to withdraw from the sanies. It dies on the spot, poisoned by its excessively high game.

What can be the meaning of this sudden corruption of the victuals, followed by the death of the Scolia, when everything appeared to have returned to its normal condition? I see only one explanation. Disturbed in its activities and diverted from its usual courses by my interference, the grub, when replaced on the wound from which I extracted it, was unable to rediscover the lode at which it was working a few minutes earlier; it thrust its way at random into the victim's entrails; and a few untimely bites extinguished the last sparks of vitality. Its confusion rendered it clumsy; and the mistake cost it its life. It dies poisoned by the rich food which, if consumed according to the rules, should have made it grow plump and lusty.

I was anxious to observe the deadly effects of a disturbed meal in another fashion. This time the victim itself shall disorder the grub's activities. The Cetonia-larva, as served up to the young Scolia by its mother, is profoundly paralysed. Its inertia is complete and so striking that it constitutes one of the leading features of this narrative. But we will not anticipate. For the moment, the thing is to substitute for this inert larva a similar larva, but one not paralysed, one very much alive. To ensure that it shall not double up and crush the grub, I confine myself to reducing it to helplessness, leaving it otherwise just as I extracted it from its burrow. I must also be careful of its legs and mandibles, the least touch of which would rip open the nurseling. With a few turns of the finest wire I fix it to a little slab of cork, with its belly in the air. Next, to provide the grub with a readymade hole, knowing that it will refuse to make one for itself, I contrive a slight incision in the skin, at the point where the Scolia lays her egg. I now place the grub upon the larva, with its head touching the bleeding wound, and lay the whole on a bed of mould in a transparent beaker protected by a pane of glass.

Unable to move, to wriggle, to scratch with its legs or snap with its mandibles, the Cetonia-larva, a new Prometheus bound, offers its defenceless flanks to the little Vulture destined to devour its entrails. Without too much hesitation, the young Scolia settles down to the wound made by my scalpel, which to the grub represents the wound whence I have just removed it. It thrusts its neck into the belly of its prey; and for a couple of days all seems to go well. Then, lo and behold, the Cetonia turns putrid and the Scolia dies, poisoned by the ptomaines of the decomposing game! As before, I see it turn brown and die on the spot, still half inside the toxic corpse.

The fatal issue of my experiment is easily explained. The Cetonia-larva is alive in every sense. True, I have, by means of bonds, suppressed its outward movements, in order to provide the nurseling with a quiet meal, devoid of danger; but it was not in my power to subdue its internal movements, the quivering of the viscera and muscles irritated by its forced immobility and by the Scolia's bites. The victim is in possession of its full power of sensation; and it expresses the pain experienced as best it may, by contractions. Embarrassed by these tremors, these twitches of suffering flesh, incommoded at every mouthful, the grub chews away at random and kills the larva almost as soon as it has started on it. In a victim paralysed by the regulation sting, the conditions would be very different. There are no external movements, nor any internal movements either, when the mandibles bite, because the victim is insensible. The grub, undisturbed in any way, is then able, with an unfaltering tooth, to pursue its scientific method of eating.

These marvellous results interested me too much not to inspire me with fresh devices when I pursued my investigations. Earlier enquiries had taught me that the larvae of the Digger-wasps are fairly indifferent to the nature of the game, though the mother always supplies them with the same diet. I had succeeded in rearing them on a great variety of prey, without paying regard to their normal fare. I shall return to this subject later, when I hope to demonstrate its great philosophical significance. Let us profit by these data and try to discover what happens when we give the Scolia food which is not properly its own.

I select from my heap of garden-mould, that inexhaustible mine, two larvae of the Rhinoceros Beetle, Oryctes nasicornis, about one-third full-grown, so that their size may not be out of proportion to the Scolia's. It is in fact almost identical with the size of the Cetonia. I paralyse one of them by giving an injection of ammonia in the nerve-centres. I make a fine incision in its belly and I place the Scolia on the opening. The dish pleases my charge; and it would be strange indeed if this were not so, considering that another Scolia-grub, the larva of the Garden Scolia, feeds on the Oryctes. The dish suits it, for before long it has burrowed halfway into the succulent paunch. This time all goes well. Will the rearing be successful? Not a bit of it! On the third day, the Oryctes decomposes and the Scolia dies. Which shall we hold responsible for the failure, myself or the grub? Myself who, perhaps too unskilfully, administered the injection of ammonia, or the grub which, a novice at dissecting a prey differing from its own, did not know how to practise its craft upon a changed victim and began to bite before the proper time?

In my uncertainty, I try again. This time I shall not interfere, so that my clumsiness cannot be to blame. As I described when speaking of the Cetonia-larva, the Oryctes-larva now lies bound, quite alive, on a strip of cork. As usual, I make a small opening in the belly, to entice the grub by means of a bleeding wound and facilitate its access. I obtain the same negative result. In a little while, the Oryctes is a noisome mass on which the nurseling lies poisoned. The failure was foreseen: to the difficulties presented by a prey unknown to my charge was added the commotion caused by the wriggling of an unparalysed animal.

We will try once more, this time with a victim paralysed not by me, an unskilled operator, but by an adept whose ability ranks so high that it is beyond discussion. Chance favours me to perfection: yesterday, in a warm sheltered corner, at the foot of a sandy bank, I discovered three cells of the Languedocian Sphex, each with its Ephippiger and the recently laid egg. This is the game I want, a corpulent prey, of a size suited to the Scolia and, what is more, in splendid condition, artistically paralysed according to rule by a master among masters.

As usual, I install my three Ephippigers in a glass jar, on a bed of mould; I remove the egg of the Sphex and on each victim, after slightly incising the skin of the belly, I place a young Scolia-grub. For three or four days my charges feed upon this game, so novel to them, without any sign of repugnance or hesitation. By the fluctuations of the digestive canal I perceive that the work of nutrition is proceeding as it should; things are happening just as if the dish were a Cetonia-larva. The change of diet, complete though it is, has in no way affected the appetite of the Scolia-grubs. But this prosperous condition does not last long. About the fourth day, a little sooner in one case, a little later in another, the three Ephippigers become putrid and the Scoliae die at the same time.

This result is eloquent. Had I left the egg of the Sphex to hatch, the larva coming out of it would have fed upon the Ephippiger; and for the hundredth time I should have witnessed an incomprehensible spectacle, that of an animal which, devoured piecemeal for nearly a fortnight, grows thin and empty, shrivels up and yet retains to the very end the freshness peculiar to living flesh. Substitute for this Sphex-larva a Scolia-larva of almost the same size; let the dish be the same though the guest is different; and healthy live flesh is promptly replaced by pestilent rotten flesh. That which under the mandibles of the Sphex would for a long while have remained wholesome food promptly becomes a poisonous liquescence under the mandibles of the Scolia.

It is impossible to explain the preservation of the victuals until finally consumed by supposing that the venom injected by the Wasp when she delivers her paralysing stings possesses antiseptic properties. The three

Ephippigers were operated on by the Sphex. Able to keep fresh under the mandibles of the Sphex-larvae, why did they promptly go bad under the mandibles of the Scolia-larvae? Any idea of an antiseptic must needs be rejected: a liquid preservative which would act in the first case could not fail to act in the second, as its virtues would not depend on the teeth of the consumer.

Those of you who are versed in the knowledge attaching to this problem, investigate, I beg you, search, sift, see if you can discover the reason why the victuals keep fresh when consumed by a Sphex, whereas they promptly become putrid when consumed by a Scolia. For me, I see only one reason; and I very much doubt whether any one can suggest another.

Both larvae practise a special art of eating, which is determined by the nature of the game. The Sphex, when sitting down to an Ephippiger, the food that has fallen to its lot, knows thoroughly how to consume it and how to preserve, to the very end, the glimmer of life which keeps it fresh; but, if it has to browse upon a Cetonia-grub, whose different structure would confuse its talents as a dissector, it would soon have nothing before it but a heap of putrescence. The Scolia, in its turn, is familiar with the method of eating the Cetonia-grub, its invariable portion; but it does not understand the art of eating the Ephippiger, though the dish is to its taste. Unable to dissect this unknown species of game, its mandibles slash away at random, killing the creature outright as soon as they take their first bites of the deeper tissues of the victim. That is the whole secret.

One more word, on which I shall enlarge in another chapter. I observe that the Scoliae to which I give

Ephippigers paralysed by the Sphex keep in excellent condition, despite the change of diet, so long as the provisions retain their freshness. They languish when the game goes high; and they die when putridity supervenes. Their death, therefore, is due not to an unaccustomed diet, but to poisoning by one or other of those terrible toxins which are engendered by animal corruption and which chemistry calls by the name of ptomaines. Therefore, notwithstanding the fatal outcome of my three attempts, I remain persuaded that the unfamiliar method of rearing would have been perfectly successful had the Ephippigers not gone bad, that is, if the Scoliae had known how to eat them according to the rules.

What a delicate and dangerous thing is the art of eating in these carnivorous larvae supplied with a single victim, which they have to spend a fortnight in consuming, on the express condition of not killing it until the very end! Could our physiological science, of which, with good reason, we are so proud, describe, without blundering, the method to be followed in the successive mouthfuls? How has a miserable grub learnt what our knowledge cannot tell us? By habit, the Darwinians will reply, who see in instinct an acquired habit.

Before deciding this serious matter, I will ask you to reflect that the first Wasp, of whatever kind, that thought of feeding her progeny on a Cetonia-grub or on any other large piece of game demanding long preservation could necessarily have left no descendants unless the art of consuming food without causing putrescence had been practised, with all its scrupulous caution, from the first generation onwards. Having as yet learnt nothing by habit or by atavistic transmission, since it was making a first beginning, the nurseling would bite into its provender at random. It would be starving, it would have no respect for its prey. It would carve its joint at random; and we have just seen the fatal consequence of an ill-directed bite. It would perish - I have just proved this in the most positive manner - it would perish, poisoned by its victim, already dead and putrid.

To prosper, it would have, although a novice, to know what was permitted and what forbidden in ransacking the creature's entrails; nor would it be enough for the larva to be approximately in possession of this difficult secret: it would be indispensable that it should possess the secret completely, for a single bite, if delivered before the right moment, would inevitably involve its own demise. The Scoliae of my experiments are not novices, far from it: they are the descendants of carvers that have practised their art since Scoliae first came into the world; nevertheless they all perish from the decomposition of the rations supplied, when I try to feed them on Ephippigers paralysed by the Sphex. Very expert in the method of attacking the Cetonia, they do not know how to set about the business of discreetly consuming a species of game new to them. All that escapes them is a few details, for the trade of an ogre fed on live flesh is familiar to them in its general features; and these unheeded details are enough to turn their food into poison. What, then, happened in the beginning, when the larva bit for the first time into a luscious victim? The inexperienced creature perished; of that there is not a shadow of doubt, unless we admit an absurdity and imagine the larva of antiquity feeding upon those terrible ptomaines which so swiftly kill its descendants today.

Nothing will ever make me admit and no unprejudiced mind can admit that what was once food has become a horrible poison. What the larva of antiquity ate was live flesh and not putrescence. Nor can it be admitted that the chances of fortune can have led at the first trial to success in a system of nourishment so full of pitfalls: fortuitous results are preposterous amid so many complications. Either the feeding is strictly methodical at the beginning, in conformity with the organic exigencies of the prey devoured, and the Wasp established her race; or else it was hesitating, without determined rules, and the Wasp left no successor. In the first case we behold innate instinct; in the second acquired habit.

A strange acquisition, truly! An acquisition presumed to be made by an impossible creature; an acquisition supposed to develop in no less impossible successors! Though the snowball, slowly rolling, at last becomes an enormous sphere, it is still necessary that the starting-point shall not have been NIL. The big ball implies the little ball, as small as you please. Now, in harking back to the origin of these acquired habits, if I interrogate the possibilities I obtain zero as the only answer. If the animal does not know its trade thoroughly, if it has to acquire something, all the more if it has to acquire everything, it perishes: that is inevitable; without the little snowball the big snowball cannot be rolled. If it has nothing to acquire, if it knows all that it needs to know, it flourishes and leaves descendants behind it. But then it possesses innate instinct, the instinct which learns nothing and forgets nothing, the instinct which is steadfast throughout time.

The building up of theories has never appealed to me: I suspect them one and all. To argue nebulously upon dubious premises likes me no better. I observe, I experiment and I let the facts speak for themselves. We have just heard these facts. Let each now decide for himself whether instinct is an innate faculty or an acquired habit.

CHAPTER 4.

THE CETONIA-LARVA.

***

The Scolia's feeding-period lasts, on the average, for a dozen days or so. By then the victuals are no more than a crumpled bag, a skin emptied of the last scrap of nutriment. A little earlier, the russet-yellow tint announces the extinction of the last spark of life in the creature that is being devoured. The empty skin is pushed back to make space; the dining-room, a shapeless cavity with crumbling walls, is tidied up a little; and the Scolia-grub sets to work on its cocoon without further delay.

The first courses form a general scaffolding, which finds a support here and there on the earthen walls, and consist of a rough, blood-red fabric. When the larva is merely laid, as required by my investigations, in a hollow made with the fingertip in the bed of mould, it is not able to spin its cocoon, for want of a ceiling to which to fasten the upper threads of its network. To weave its cocoon, every spinning larva is compelled to isolate itself in a hammock slung in an openwork enclosure, which enables it to distribute its thread uniformly in all directions. If there be no ceiling, the upper part of the cocoon cannot be fashioned, because the worker lacks the necessary points of support. Under these conditions my Scolia-grubs contrive at most to upholster their little pit with a thick down of reddish silk. Discouraged by futile endeavours, some of them die. It is as if they had been killed by the silk which they omit to disgorge because they are unable to make the right use of it. This, if we were not watchful, would be a very frequent cause of failure in our attempts at artificial rearing. But, once the danger has been perceived, the remedy is simple. I make a ceiling over the cavity by laying a short strip of paper above it. If I want to see how matters are progressing, I bend the strip into a semicircle, into a half-cylinder with open ends. Those who wish to play the breeder for themselves will be able to profit by these little practical details.

In twenty-four hours the cocoon is finished; at least, it no longer allows us to see the grub, which is doubtless making the walls of its dwelling still thicker. At first the cocoon is a vivid red; later it changes to a light chestnut-brown. Its form is that of an ellipsoid, with a major axis 26 millimetres in length, while the minor axis measures 11 millimetres. These dimensions, which incidentally are inclined to vary slightly, are those of the female cocoons. In the other sex they are smaller and may measure as little as 17 millimetres in length by 7 millimetres in width.

The two ends of the ellipsoid have the same form, so much so that it is only thanks to an individual peculiarity, independent of the shape, that we can tell the cephalic from the anal extremity. The cephalic pole is flexible and yields to the pressure of my tweezers; the anal pole is hard and unyielding. The wrapper is double, as in the cocoons of the Sphex. The outer envelope, consisting of pure silk, is thin, flexible and offers little resistance. It is closely superimposed upon the inner envelope and is easily separated from it everywhere, except at the anal end, where it adheres to the second envelope. The adhesion of the two wrappers at one end and the non-adhesion at the other are the cause of the differences which the tweezers reveal when pinching the two ends of the cocoon.

The inner envelope is firm, elastic, rigid and, to a certain point, brittle. I do not hesitate to look upon it as consisting of a silken tissue which the larva, towards the end of its task, has steeped thoroughly in a sort of varnish prepared not by the silk-glands but by the stomach. The cocoons of the Sphex have already shown us a similar varnish. This product of the chylific ventricle is chestnut-brown. It is this which, saturating the thickness of the tissue, effaces the bright red of the beginning and replaces it by a brown tint. It is this again which, disgorged more profusely at the lower end of the cocoon, glues the two wrappers together at that point.

The perfect insect is hatched at the beginning of July. The emergence takes place without any violent effraction, without any ragged rents. A clean, circular fissure appears at some distance from the top; and the cephalic end is detached all of a piece, as a loose lid might be. It is as though the recluse had only to raise a cover by butting it with her head, so exact is the line of division, at least as regards the inner envelope, the stronger and more important of the two. As for the outer wrapper, its lack of resistance enables it to yield without difficulty when the other gives way.

I cannot quite make out by what knack the Wasp contrives to detach the cap of the inner shell with such accuracy. Is it the art practised by the tailor when cutting his stuff, with mandibles taking the place of scissors? I hardly venture to admit as much: the tissue is so tough and the circle of division so precise.

The mandibles are not sharp enough to cut without leaving a ragged edge; and then what geometrical certainty they would need for an operation so perfect that it might well have been performed with the compasses!

I suspect therefore that the Scolia first fashions the outer sac in accordance with the usual method, that is, by distributing the silk uniformly, without any special preparation of one part of the wall more than of another, and that it afterwards changes its method of weaving in order to attend to the main work, the inner shell. In this it apparently imitates the Bembex, which weaves a sort of eel-trap, whose ample mesh allows it to gather grains of sand outside and encrust them one by one in the silky network, and completes the performance with a cap fitting the entrance to the trap. This provides a circular line of least resistance, along which the casket breaks open afterwards. If the Scolia really works in the same manner, everything is explained: the eel-trap, while still open, enables it to soak with varnish both the inside and the outside of the inner shell, which has to acquire the consistency of parchment; lastly, the cap which completes and closes the structure leaves for the future a circular line capable of splitting easily and neatly.

This is enough on the subject of the Scolia-grub. Let us go back to its provender, of whose remarkable structure we as yet know nothing. In order that it may be consumed with the delicate anatomical discretion imposed by the necessity of having fresh food to the last, the Cetonia-grub must be plunged into a state of absolute immobility: any twitchings on its part - as the experiments which I have undertaken go to prove - would discourage our nibbling larva and impede the work of carving, which has to be effected with so much circumspection. It is not enough for the victim to be unable to move from place to place beneath the soil: in addition to this, the contractible power in its sturdy muscular organism must be suppressed.

In its normal state, this larva, at the very least disturbance, curls itself up, almost as the Hedgehog does; and the two halves of the ventral surface are laid one against the other. You are quite surprised at the strength which the creature displays in keeping itself thus contracted. If you try to unroll it, your fingers encounter a resistance far greater than the size of the animal would have caused you to suspect. To overcome the resistance of this sort of spring coiled upon itself, you have to force it, so much so that you are afraid, if you persist, of seeing the indomitable spiral suddenly burst and shoot forth its entrails.

A similar muscular energy is found in the larvae of the Oryctes, the Anoxia, the Cockchafer. Weighed down by a heavy belly and living underground, where they feed either on leaf-mould or on roots, these larvae all possess the vigorous constitution needed to drag their corpulence through a resisting medium. All of them also roll themselves into a hook which is not straightened without an effort.

Now what would become of the egg and the newborn grub of the Scoliae, fixed under the belly, at the centre of the Cetonia's spiral, or inside the hook of the Oryctes or the Anoxia? They would be crushed between the jaws of the living vice. It is essential that the arc should slacken and the hook unbend, without the least possibility of their returning to a state of tension. Indeed, the wellbeing of the Scoliae demands something more: those powerful bodies must not retain even the power to quiver, lest they derange a method of feeding which has to be conducted with the greatest caution.

The Cetonia-grub to which the Two-banded Scolia's egg is fastened fulfils the required conditions admirably. It is lying on its back, in the midst of the mould, with its belly fully extended. Long accustomed though I be to this spectacle of victims paralysed by the sting of the Hunting Wasp, I cannot suppress my astonishment at the profound immobility of the prey before my eyes. In the other victims with flexible skins, Caterpillars, Crickets, Mantes, Ephippigers, I perceived at least some pulsations of the abdomen, a few feeble contortions under the stimulus of a needle. There is nothing of the sort here, nothing but absolute inertia, except in the head, where I see, from time to time, the mouthparts open and close, the palpi give a tremor, the short antennae sway to and fro. A prick with the point of a needle causes no contraction, no matter what the spot pricked. Though I stab it through and through, the creature does not stir, be it ever so little. A corpse is not more inert. Never, since my remotest investigations, have I witnessed so profound a paralysis. I have seen many wonders due to the surgical talent of the Wasp; but today's marvel surpasses them all.

I am doubly surprised when I consider the unfavourable conditions under which the Scolia operates. The other paralysers work in the open air, in the full light of day. There is nothing to hinder them. They enjoy full liberty of action in seizing the prey, holding it in position and sacrificing it; they are able to see the victim and to parry its means of defence, to avoid its spears, its pincers. The spot or spots to be attained are within their reach; they drive the dagger in without let or hindrance.

What difficulties, on the other hand, await the Scolia! She hunts underground, in the blackest darkness. Her movements are laboured and uncertain, owing to the mould, which is continually giving way all round her; she cannot keep her eyes on the terrible mandibles, which are capable of cutting her body in two with a single bite. Moreover, the Cetonia-grub, perceiving that the enemy is approaching, assumes its defensive posture, rolls itself up and makes a shield for its only vulnerable part, the ventral surface, with its convex back. No, it cannot be an easy operation to subdue the powerful larva in its underground retreat and to stab with the precision which immediate paralysis requires.

We wish that we might witness the struggle between the two adversaries and see at first hand what happens, but we cannot hope to succeed. It all takes place in the mysterious darkness of the soil; in broad daylight, the attack would not be delivered, for the victim must remain where it is and then and there receive the egg, which is unable to thrive and develop except under the warm cover of vegetable mould. If direct observation is impracticable, we can at least foresee the main outlines of the drama by allowing ourselves to be guided by the warlike manoeuvres of other burrowers.

I picture things thus: digging and rummaging through the heap of mould, guided perhaps by that singular sensibility of the antennae which enables the Hairy Ammophila to discover the Grey Worm underground, the Scolia ends by finding a Cetonia-larva, a good plump one, in the pink of condition, having reached its full growth, just what the grub which is to feed on it requires. Forthwith, the assaulted victim, contracting desperately, rolls itself into a ball. The other seizes it by the skin of the neck. To unroll it is impossible to the insect, for I myself have some trouble in doing so. One single point is accessible to the sting: the under part of the head, or rather of the first segments, which are placed outside the coil, so that the grub's hard cranium makes a rampart for the hinder extremity, which is less well defended. Here the Wasp's sting enters and here only can it enter, within a narrowly circumscribed area. One stab only of the lancet is given at this point, one only because there is no room for more; and this is enough: the larva is absolutely paralysed.

The nervous functions are abolished instantly; the muscular contractions cease; and the animal uncoils like a broken spring. Henceforth motionless, it lies on its back, its ventral surface fully exposed from end to end. On the median line of this surface, towards the rear, near the brown patch due to the alimentary broth contained in the intestine, the Scolia lays her egg and without more ado, leaves everything lying on the actual spot where the murder was committed, in order to go in search of another victim.

This is how the deed must be done: the results prove it emphatically. But then the Cetonia-grub must possess a very exceptional structure in its nervous organization. The larva's violent contraction leaves but a single point of attack open to the sting, the under part of the neck, which is doubtless uncovered when the victim tries to defend itself with its mandibles; and yet a stab in this one point produces the most thorough paralysis that I have ever seen. It is the general rule that larvae possess a centre of innervation for each segment. This is so in particular with the Grey Worm, the sacrificial victim of the Hairy Ammophila. The Wasp is acquainted with this anatomical secret: she stabs the caterpillar again and again, from end to end, segment by segment, ganglion by ganglion. With such an organization the Cetonia-grub, unconquerably coiled upon itself would defy the paralyser's surgical skill.

If the first ganglion were wounded, the others would remain uninjured; and the powerful body, actuated by these last, would lose none of its powers of contraction. Woe then to the egg, to the young grub held fast in its embrace! And how insurmountable would be the difficulties if the Scolia, working in the profound darkness amid the crumbling soil and confronted by a terrible pair of mandibles, had to stab each segment in turn with her sting, with the certainty of method displayed by the Ammophila! The delicate operation is possible in the open air, where nothing stands in the way, in broad daylight, where the sight guides the scalpel, and with a patient which can always be released if it becomes dangerous. But in the dark, underground, amidst the ruins of a ceiling which crumbles in consequence of the conflict and at close quarters with an opponent greatly her superior in strength, how is the Scolia to guide her sting with the accuracy that is essential if the stabs are to be repeated?

So profound a paralysis; the difficulty of vivisection underground; the desperate coiling of the victim: all these things tell me that the Cetonia-grub, as regards its nervous system, must possess a structure peculiar to itself. The whole of the ganglia must be concentrated in a limited area in the first segments, almost under the neck. I see this as clearly as though it had been revealed to me by a postmortem dissection.

Never was anatomical forecast more fully confirmed by direct examination. After forty-eight hours in benzine, which dissolves the fat and renders the nervous system more plainly visible, the Cetonia-grub is subjected to dissection. Those of my readers who are familiar with these investigations will understand my delight. What a clever school is the Scolia's! It is just as I thought! Admirable! The thoracic and abdominal ganglia are gathered into a single nervous mass, situated within the quadrilateral bounded by the four hinder legs, which legs are very near the head. It is a tiny, dull-white cylinder, about three millimetres long by half a millimetre wide. This is the organ which the Scolia's sting must attack in order to secure the paralysis of the whole body, excepting the head, which is provided with special ganglia. From it run numbers of filaments which actuate the feet and the powerful muscular layer which is the creature's essential motor organ. When examined merely through the pocket-lens, this cylinder appears to be slightly furrowed transversely, a proof of its complex structure. Under the microscope, it is seen to be formed by the close juxtaposition, the welding, end to end, of the ganglia, which can be distinguished one from the other by a slight intermediate groove. The bulkiest are the first, the fourth and the tenth, or last; these are all very nearly of equal size. The rest are barely half or even a third as large as those mentioned.

The Interrupted Scolia experiences the same hunting and surgical difficulties when she attacks, in the crumbling, sandy soil, the larvae of the Shaggy Anoxia or of the Morning Anoxia, according to the district; and these difficulties, if they are to be overcome, demand in the victim a concentrated nervous system, like the Cetonia's. Such is my logical conviction before making my examination; such also is the result of direct observation. When subjected to the scalpel, the larva of the Morning Anoxia shows me its centres of innervation for the thorax and the abdomen, gathered into a short cylinder, which, placed very far forward, almost immediately after the head, does not run back beyond the level of the second pair of legs. The vulnerable point is thus easily accessible to the sting, despite the creature's posture of defence, in which it contracts and coils up. In this cylinder I recognize eleven ganglia, one more than in the Cetonia. The first three, or thoracic, ganglia are plainly distinguishable from one another, although they are set very close together; the rest are all in contact. The largest are the three thoracic ganglia and the eleventh.

After ascertaining these facts, I remembered Swammerdam's investigations into the grub of the Monoceros, our Oryctes nasicornis. I chanced to possess an abridgement of the "Biblia naturae," the masterly work of the father of insect anatomy. I consulted the venerable volume. It informed me that the learned Dutchman had been struck, long before I was, by an anatomical peculiarity similar to that which the larvae of the Cetoniae and Anoxiae had shown me in their nerve-centres. Having observed in the silkworm a nervous system formed of ganglia distinct one from the other, he was quite surprised to find that, in the grub of the Oryctes, the same system was concentrated into a short chain of ganglia in juxtaposition. His was the surprise of the anatomist who, studying the organ qua organ, sees for the first time an unusual conformation. Mine was of a different nature: I was amazed to see the precision with which the paralysis of the victim sacrificed by the Scolia, a paralysis so profound in spite of the difficulties of an underground operation, had guided my forecast as to structure when, anticipating the dissection, I declared in favour of an exceptional concentration of the nervous system. Physiology perceived what anatomy had not yet revealed, at all events to my eyes, for since then, on dipping into my books, I have learnt that these anatomical peculiarities, which were then so new to me, are now within the domain of current science. We know that, in the Scarabaeidae, both the larva and the perfect insect are endowed with a concentrated nervous system.

The Garden Scolia attacks Oryctes nasicornis; the Two-banded Scolia the Cetonia; the Interrupted Scolia the Anoxia. All three operate below ground, under the most unfavourable conditions; and all three have for their victim a larva of one of the Scarabaeidae, which, thanks to the exceptional arrangement of its nerve-centres, lends itself, alone of all larvae, to the Wasp's successful enterprises. In the presence of this underground game, so greatly varied in size and shape and yet so judiciously selected to facilitate paralysis, I do not hesitate to generalize and I accept, as the ration of the other Scoliae, larvae of Lamellicorns whose species will be determined by future observation. Perhaps one of them will be found to give chase to the terrible enemy of my crops, the voracious White Worm, the grub of the Cockchafer; perhaps the Hemorrhoidal Scolia, rivalling in size the Garden Scolia and like her, no doubt, requiring a copious diet, will be entered in the insects' "Who's Who" as the destroyer of the Pine-chafer, that magnificent Beetle, flecked with white upon a black or brown ground, who of an evening, during the summer solstice, browses on the foliage of the fir-trees. Though unable to speak with certainty or precision, I am inclined to look upon these devourers of Scarabaeus-grubs as valiant agricultural auxiliaries.

The Cetonia-larva has figured hitherto only in its quality of a paralysed victim. We will now consider it in its normal state. With its convex back and its almost flat ventral surface, the creature is like a semi-cylinder in shape, fuller in the hinder portion. On the back, each of the segments, except the last, or anal, segment, puckers into three thick pads, bristling with stiff, tawny hairs. The anal segment, much wider than the rest, is rounded at the end and coloured a deep brown by the contents of the intestine, which show through the translucent skin; it bristles with hairs like the other segments, but is level, without pads. On the ventral surface, the segments have no creases; and the hairs, though abundant, are rather less so than on the back. The legs, which are quite well-formed, are short and feeble in comparison with the animal's size. The head has a strong, horny cap for a cranium. The mandibles are powerful, with bevelled tips and three or four teeth on the edge of the bevel.

Its mode of locomotion marks it as an idiosyncratic, exceptional, fantastic creature, having no fellow, that I know of, in the insect world. Though endowed with legs - a trifle short, it is true, but after all as good as those of a host of other larvae - it never uses them for walking. It progresses on its back, always on its back, never otherwise. By means of wriggling movements and the purchase afforded by the dorsal bristles, it makes its way belly upwards, with its legs kicking the empty air. The spectator to whom these topsy-turvy gymnastics are a novelty thinks at first that the creature must have had a fright of some sort and that it is struggling as best it can in the face of danger. He puts it back on its belly; he lays it on its side. Nothing is of any use; it obstinately turns over and resumes its dorsal progress. That is its manner of travelling over a flat surface; it has no other.

This reversal of the usual mode of walking is so peculiar to the Cetonia-larva that it is enough in itself to reveal the grub's identity to the least expert eyes. Dig into the vegetable mould formed by the decayed wood in the hollow trunks of old willow-trees, search at the foot of rotten stumps or in heaps of compost; and, if you come upon a plumpish grub moving along on its back, there is no room for doubt: your discovery is a Cetonia-larva.

This topsy-turvy progress is fairly swift and is not less in speed to that of an equally fat grub travelling on its legs. It would even be greater on a polished surface, where walking on foot is hampered by incessant slips, whereas the numerous hairs of the dorsal pads find the necessary support by multiplying the points of contact. On polished wood, on a sheet of paper and even on a strip of glass, I see my grubs moving from point to point with the same ease as on a surface of garden mould. In the space of one minute, on the wood of my table, they cover a distance of eight inches. The pace is no swifter on a horizontal bed of sifted mould. A strip of glass reduces the distance covered by one half. The slippery surface only half paralyses this strange method of locomotion.

We will now place side by side with the Cetonia-grub the larva of the Morning Anoxia, the prey of the Interrupted Scolia. It is very like the larva of the Common

Cockchafer. It is a fat, potbellied grub, with a thick, red cap on its head and armed with strong, black mandibles, which are powerful implements for digging and cutting through roots. The legs are sturdy and end in a hooked nail. The creature has a long, heavy, brown paunch. When placed on the table, it lies on its side; it struggles without being able to advance or even to remain on its belly or back. In its usual posture it is curled up into a narrow hook. I have never seen it straighten itself completely; the bulky abdomen prevents it. When placed on a surface of moist sand, the ventripotent creature is no better able to shift its position: curved into a fishhook, it lies on its side.

To dig into the earth and bury itself, it uses the fore-edge of its head, a sort of weeding-hoe with the two mandibles for points. The legs take part in this work, but far less effectually. In this way it contrives to dig itself a shallow pit. Then, bracing itself against the wall of the pit, with the aid of wriggling movements which are favoured by the short, stiff hairs bristling all over its body, the grub changes its position and plunges into the sand, but still with difficulty.

Apart from a few details, which are of no importance here, we may repeat this sketch of the Anoxia-grub and we shall have, if the size be at least quadrupled, a picture of the larva of Oryctes nasicornis, the monstrous prey of the Garden Scolia. Its general appearance is the same: there is the same exaggeration of the belly; the same hook-like curve; the same incapacity for standing on its legs. And as much may be said of the larva of Scarabaeus pentodon, a fellow-boarder of the Oryctes and the Cetonia.

CHAPTER 5.

THE PROBLEM OF THE SCOLIAE.

***

Now that all the facts have been set forth, it is time to collate them. We already know that the Beetle-hunters, the Cerceres, prey exclusively on the Weevils and the Buprestes, that is, on the families whose nervous system presents a degree of concentration which may be compared with that of the Scolia's victims. Those predatory insects, working in the open air, are exempt from the difficulties which their emulators, working underground, have to overcome. Their movements are free and are directed by the sense of sight; but their surgery is confronted in another respect with a most arduous problem.

The victim, a Beetle, is covered at all points with a suit of armour which the sting is unable to penetrate. The joints alone will allow the poisoned lancet to pass. Those of the legs do not in any way comply with the conditions imposed: the result of stinging them would be merely a partial disorder which far from subduing the insect, would render it more dangerous by irritating it yet further. A sting in the joint of the neck is not admissible: it would injure the cervical ganglia and lead to death, followed by putrefaction. There remains only the joint between the corselet and the abdomen.

The sting, in entering here, has to abolish all movement with a single stab, for any movement would imperil the rearing of the larva. The success of the paralysis, therefore, demands that the motor ganglia, at least the three thoracic ganglia, shall be packed in close contact opposite this point. This determines the selection of Weevils and Buprestes, both of which are so strongly armoured.

But where the prey has only a soft skin, incapable of stopping the sting, the concentrated nervous system is no longer necessary, for the operator, versed in the anatomical secrets of her victim, knows to perfection where the centres of innervation lie; and she wounds them one after another, if need be from the first to the last. Thus do the Ammophilae go to work when dealing with their caterpillars and the Sphex-wasps when dealing with their Locusts, Ephippigers and Crickets.

With the Scoliae we come once again to a soft prey, with a skin penetrable by the sting no matter where it be attacked. Will the tactics of the caterpillar-hunters, who stab and stab again, be repeated here? No, for the difficulty of movement under ground prohibits so complicated an operation. Only the tactics of the paralysers of armour-clad insects are practicable now, for, since there is but one thrust of the dagger, the feat of surgery is reduced to its simplest terms, a necessary consequence of the difficulties of an underground operation. The Scoliae, then, whose destiny it is to hunt and paralyse under the soil the victuals for their family, require a prey made highly vulnerable by the close assemblage of the nerve-centres, as are the Weevils and Buprestes of the Cerceres; and this is why it has fallen to their lot to share among them the larvae of the Scarabaeidae.

Before they obtained their allotted portion, so closely restricted and so judiciously selected; before they discovered the precise and almost mathematical point at which the sting must enter to produce a sudden and a lasting immobility; before they learnt how to consume, without incurring the risk of putrefaction, so corpulent a prey: in brief, before they combined these three conditions of success, what did the Scoliae do?

The Darwinian school will reply that they were hesitating, essaying, experimenting. A long series of blind gropings eventually hit upon the most favourable combination, a combination henceforth to be perpetuated by hereditary transmission. The skilful co-ordination between the end and the means was originally the result of an accident.

Chance! A convenient refuge! I shrug my shoulders when I hear it invoked to explain the genesis of an instinct so complex as that of the Scoliae. In the beginning, you say, the creature gropes and feels its way; there is nothing settled about its preferences. To feed its carnivorous larvae it levies tribute on every species of game which is not too much for the huntress' power or the nurseling's appetite; its descendants try now this, now that, now something else, at random, until the accumulated centuries lead to the selection which best suits the race. Then habit grows fixed and becomes instinct.

Very well. Let us agree that the Scolia of antiquity sought a different prey from that adopted by the modern huntress. If the family throve upon a diet now discontinued, we fail to see that the descendants had any reason to change it: animals have not the gastronomic fancies of an epicure whom satiety makes difficult to please. Because the race did well upon this fare, it became habitual; and instinct became differently fixed from what it is today. If, on the other hand, the original food was unsuitable, the existence of the family was jeopardized; and any attempt at future improvement became impossible, because an unhappily inspired mother would leave no heirs.

To escape falling into this twofold trap, the theorists will reply that the Scoliae are descended from a precursor, an indeterminate creature, of changeable habits and changing form, modifying itself in accordance with its environment and with the regional and climatic conditions and branching out into races each of which has become a species with the attributes which distinguish it today. The precursor is the deus ex machina of evolution. When the difficulty becomes altogether too importunate, quick, a precursor, to fill up the gaps, quick, an imaginary creature, the nebulous plaything of the mind! This is seeking to lighten the darkness with a still deeper obscurity; to illumine the day by piling cloud upon cloud. Precursors are easier to find than sound arguments. Nevertheless, let us put the precursor of the Scoliae to the test.

What did she do? Being capable of everything, she did a bit of everything. Among its descendants were innovators who developed a taste for tunnelling in sand and vegetable mould. There they encountered the larvae of the Cetonia, the Oryctes, the Anoxia, succulent morsels on which to rear their families. By degrees the indeterminate Wasp adopted the sturdy proportions demanded by underground labour. By degrees she learnt to stab her plump neighbours in scientific fashion; by degrees she acquired the difficult art of consuming her prey without killing it; at length, by degrees, aided by the richness of her diet, she became the powerful Scolia with whom we are familiar. Having reached this point, the species assumes a permanent form, as does its instinct.

Here we have a multiplicity of stages, all of the slowest, all of the most incredible nature, whereas the Wasp cannot found a race except on the express condition of complete success from the first attempt. We will not insist further upon the insurmountable objection; we will admit that, amid so many unfavourable chances, a few favoured individuals survive, becoming more and more numerous from one generation to the next, in proportion as the dangerous art of rearing the young is perfected. Slight variations in one and the same direction form a definite whole; and at long last the ancient precursor has become the Scolia of our own times.

By the aid of a vague phraseology which juggles with the secret of the centuries and the unknown things of life, it is easy to build up a theory in which our mental sloth delights, after being discouraged by difficult researches whose final result is doubt rather than positive statement. But if, so far from being satisfied with hazy generalities and adopting as current coin the terms consecrated by fashion, we have the perseverance to explore the truth as far as lies in our power, the aspect of things will undergo a great change and we shall discover that they are far less simple than our overprecipitate views declared them to be. Generalization is certainly a most valuable instrument: science indeed exists only by virtue of it. Let us none the less beware of generalizations which are not based upon very firm and manifold foundations.

When these foundations are lacking, the child is the great generalizer. For him, the feathered world consists merely of birds; the race of reptiles merely of snakes, the only difference being that some are big and some are little. Knowing nothing, he generalizes in the highest degree; he simplifies, in his inability to perceive the complex. Later he will learn that the Sparrow is not the Bullfinch, that the Linnet is not the Greenfinch; he will particularize and to a greater degree each day, as his faculty of observation becomes more fully trained. In the beginning he saw nothing but resemblances; he now sees differences, but still not plainly enough to avoid incongruous comparisons.

In his adult years he will almost to a certainty commit zoological blunders similar to those which my gardener retails to me. Favier, an old soldier, has never opened a book, for the best of reasons. He barely knows how to cipher: arithmetic rather than reading is forced upon us by the brutalities of life. Having followed the flag over three-quarters of the globe, he has an open mind and a memory crammed with reminiscences, which does not prevent him, when we chat about animals, from making the most crazy assertions. For him the Bat is a Rat that has grown wings; the Cuckoo is a Sparrow-hawk retired from business; the Slug is a Snail who has lost his shell with the advance of years; the Nightjar, the Chaoucho-grapaou, as he calls her, is an elderly Toad, who, becoming enamoured of milk-food, has grown feathers, so that she may enter the byres and milk the Goats. It is impossible to drive these fantastic ideas out of his head. Favier himself, as will be seen, is an evolutionist after his own fashion, an evolutionist of a very daring type. In accounting for the origin of animals nothing gives him pause. He has a reply to everything: "this" comes from "that." If you ask him why, he answers:

"Look at the resemblance!"

Shall we reproach him with these insanities, when we hear another, misled by the Monkey's build, acclaim the Pithecanthropus as man's precursor? Shall we reject the metamorphosis of the Chaoucho-grapaou, when people tell us in all seriousness that, in the present stage of scientific knowledge, it is absolutely proved that man is descended from some rough-hewn Ape? Of the two transformations, Favier's strikes me as the more credible. A painter of my acquaintance, a brother of the great composer Felicien David (Felicien Cesar David, favoured me one day with his reflections on the human structure:

"Ve, moun bel ami," he said. "Ve, l'home a lou dintre d'un por et lou defero d'uno mounino." "See, my dear friend, see: man has the inside of a pig and the outside of a monkey."

I recommend the painter's aphorism to those who might like to discover man's origin in the Hog when the Ape has gone out of fashion. According to David, descent is proved by internal resemblances:

"L'home a lou dintre d'un por."

The inventory of precursory types sees nothing but organic resemblances and disdains the differences of aptitude. By consulting only the bones, the vertebrae, the hair, the nervures of the wings, the joints of the antennae, the imagination may build up any sort of genealogical tree that will fit with our theories of classification, for, when all is said, the animal, in its widest generalization, is represented by a digestive tube. With this common factor, the way lies open to every kind of error. A machine is judged not by this or that train of wheels, but by the nature of the work accomplished. The monumental roasting-jack of a waggoners' inn and a Breguet chronometer both have trains of cogwheels geared in almost a similar fashion. Are we to class the two mechanisms together? Shall we forget that the one turns a shoulder of mutton before the hearth, while the other divides time into seconds?

In the same way, the organic scaffolding is dominated from on high by the aptitudes of the animal, especially that superior characteristic, the psychical aptitudes. That the Chimpanzee and the hideous Gorilla possess close resemblances of structure to our own is obvious. But let us for a moment consider their aptitudes. What differences, what a dividing gulf! Without exalting ourselves as high as the famous reed of which Pascal speaks, that reed which, in its weakness, by the mere fact that it knows itself to be crushed, is superior to the world that crushes it, we may at least ask to be shown, somewhere, an animal making an implement, which will multiply its skill and its strength, or taking possession of fire, the primordial element of progress. Master of implements and of fire! These two aptitudes, simple though they be, characterize man better than the number of his vertebrae and his molars.

You tell us that man, at first a hairy brute, walking on all fours, has risen on his hind-legs and shed his fur; and you complacently demonstrate how the elimination of the hairy pelt was effected. Instead of bolstering up a theory with a handful of fluff gained or lost, it would perhaps be better to settle how the original brute became the possessor of implements and fire. Aptitudes are more important than hair; and you neglect them because it is there that the insurmountable difficulty really resides. See how the great master of evolution hesitates and stammers when he tries, by fair means or foul, to fit instinct into the mould of his formulae. It is not so easy to handle as the colour of the pelt, the length of the tail, the ear that droops or stands erect. Yes, our master well knows that this is where the shoe pinches! Instinct escapes him and brings his theory crumbling to the ground.

Let us return to what the Scoliae teach us on this question, which incidentally touches on our own origin. In conformity with the Darwinian ideas, we have accepted an unknown precursor, who by dint of repeated experiment, adopted as the victuals to be hoarded the larvae of the Scarabaeidae. This precursor, modified by varying circumstances, is supposed to have subdivided herself into ramifications, one of which, digging into vegetable mould and preferring the Cetonia to any other game inhabiting the same heap, became the Two-banded Scolia; another, also addicted to exploring the soil, but selecting the Oryctes, left as its descendant the Garden Scolia; and a third, establishing itself in sandy ground, where it found the Anoxia, was the ancestress of the Interrupted Scolia. To these three ramifications we must beyond a doubt add others which complete the series of the Scolia. As their habits are known to me only by analogy, I confine myself to mentioning them.

The three species at least, therefore, with which I am familiar would appear to be derived from a common precursor. To traverse the distance from the starting-point to the goal, all three have had to contend with difficulties, which are extremely grave if considered one by one and are aggravated even more by this circumstance, that the overcoming of one would lead to nothing unless the others were surmounted as successfully. Success, then, is contingent upon a series of conditions, each one of which offers almost no chance of victory, so that the fulfilment of them all becomes a mathematical absurdity if we are to invoke accident alone.

And, in the first place, how was it that the Scolia of antiquity, having to provide rations for her carnivorous family, adopted for her prey only those larvae which, owing to the concentration of their nervous systems, form so remarkable and so rare an exception in the insect order? What chance would hazard offer her of obtaining this prey, the most suitable of all because the most vulnerable? The chance represented by unity compared with the indefinite number of entomological species. The odds are as one to immensity.

Let us continue. The larva of the Scarabaeid is snapped up underground, for the first time. The victim protests, defends itself after its fashion, coils itself up and presents to the sting on every side a surface on which a wound entails no serious danger. And yet the Wasp, an absolute novice, has to select, for the thrust of its poisoned weapon, one single point, narrowly restricted and hidden in the folds of the larva's body. If she miscalculates, she may be killed: the larva, irritated by the smarting puncture, is strong enough to disembowel her with the tusks of its mandibles. If she escapes the danger, she will nevertheless perish without leaving any offspring, since the necessary provisions will be lacking. Salvation for herself and her race depends on this: whether at the first thrust she is able to reach the little nervous plexus which measures barely one-fiftieth of an inch in width.

What chance has she of plunging her lancet into it, if there is nothing to guide her? The chance represented by unity compared with the number of points composing the victim's body. The odds are as one against immensity.

Let us proceed still further. The sting has reached the mark; the fat grub is deprived of movement. At what spots should the egg now be laid? In front, behind, on the sides, the back or the belly? The choice is not a matter of indifference. The young grub will pierce the skin of its provender at the very spot on which the egg was fixed; and, once an opening is made, it will go ahead without hesitation. If this point of attack is ill-chosen, the nurseling runs the risk of presently finding under its mandibles some essential organ, which should have been respected until the end in order to keep the victuals fresh. Remember how difficult it is to complete the rearing when the tiny larva is moved from the place chosen by the mother. The game promptly becomes putrid and the Scolia dies.

It is impossible for me to state the precise motives which lead to the adoption of the spot on which the egg is laid; I can perceive general reasons, but the details escape me, as I am not well enough versed in the more delicate questions of anatomy and entomological physiology. What I do know with absolute certainty is that the same spot is invariably chosen for laying the egg. With not a single exception, on all the victims extracted from the heap of garden mould – and they are numerous – the egg is fixed behind the ventral surface, on the verge of the brown patch formed by the contents of the digestive system.

If there be nothing to guide her, what chance has the mother of gluing her egg to this point, which is always the same because it is that most favourable to successful rearing? A very small point, represented by the ratio of two or three square millimetres to the entire surface of the victim's body.

Is this all? Not yet. The grub is hatched; it pierces the belly of the Cetonia-larva at the requisite point; it plunges its long neck into the entrails, ransacking them and filling itself to repletion. If it bite at random, if it have no other guide in the selection of tit-bits than the preference of the moment and the violence of an imperious appetite, it will infallibly incur the danger of being poisoned by putrid food, for the victim, if wounded in those organs which preserve a remnant of life in it, will die for good and all at the first mouthfuls.

The ample joint must be consumed with prudent skill: this part must be eaten before that and, after that, some other portion, always according to method, until the time approaches for the last bites. This marks the end of life for the Cetonia, but it also marks the end of the Scolia's feasting. If the grub be a novice in the art of eating, if no special instinct guide its mandibles in the belly of the prey, what chance has it of completing its perilous meal? As much as a starving Wolf would have of daintily dissecting his Sheep, when he tears at her gluttonously, rends her into shreds and gulps them down.

These four conditions of success, with chance so near to zero in each case, must all be realized together, or the grub will never be reared. The Scolia may have captured a larva with close-packed nerve-centres, a Cetonia-grub, for instance; but this will go for nothing unless she direct her sting towards the only vulnerable point. She may know the whole secret of the art of stabbing her victim, but this means nothing if she does not know where to fasten her egg. The suitable spot may be found, but all the foregoing will be useless if the grub be not versed in the method to be followed in devouring its prey while keeping it alive. It is all or nothing.

Who would venture to calculate the final chance on which the future of the Scolia, or of her precursor, is based, that complex chance whose factors are four infinitely improbable occurrences, one might almost say four impossibilities? And such a conjunction is supposed to be a fortuitous result, to which the present instinct is due! Come, come!

From another point of view again, the Darwinian theory is at variance with the Scoliae and their prey. In the heap of garden mould which I exploited in order to write this record, three kinds of larvae dwell together, belonging to the Scarabaeid group: the Cetonia, the Oryctes and Scarabeus pentodon. Their internal structure is very nearly similar; their food is the same, consisting of decomposing vegetable matter; their habits are identical: they live underground in tunnels which are frequently renewed; they make a rough egg-shaped cocoon of earthy materials. Environment, diet, industry and internal structure are all similar; and yet one of these three larvae, the Cetonia's, reveals a most singular dissimilarity from its fellow-trenchermen: alone among the Scarabaeidae and, more than that, alone in all the immense order of insects, it walks upon its back.

If the differences were a matter of a few petty structural details, falling within the finical department of the classifier, we might pass them over without hesitation; but a creature that turns itself upside down in order to walk with its belly in the air and never adopts any other method of locomotion, though it possesses legs and good legs at that, assuredly deserves examination. How did the animal acquire its fantastic mode of progress and why does it think fit to walk in a fashion the exact contrary of that adopted by other beasts?

To these questions the science now in fashion always has a reply ready: adaptation to environment. The Cetonia-larva lives in crumbling galleries which it bores in the depths of the soil. Like the sweep who obtains a purchase with his back, loins and knees to hoist himself up the narrow passage of a chimney, it gathers itself up, applies the tip of its belly to one wall of its gallery and its sturdy back to another; and the combined effort of these two levers results in moving it forward. The legs, which are used very little, indeed hardly at all, waste away and tend to disappear, as does any organ which is left unemployed; the back, on the other hand, the principal motive agent, grows stronger, is furrowed with powerful folds and bristles with grappling-hooks or hairs; and gradually, by adaptation to its environment, the creature loses the art of walking, which it does not practise, and replaces it by that of crawling on its back, a form of progress better suited to underground corridors.

So far so good. But now tell me, if you please, why the larvae of the Oryctes and the Scarabaeus, living in vegetable mould, the larva of the Anoxia, dwelling in the sand, and the larva of the Cockchafer in our cultivated fields have not also acquired the faculty of walking on their backs? In their galleries they follow the chimneysweep's methods quite as cleverly as the Cetonia-grub; to move forward they make valiant use of their backs without yet having come to ambling with their bellies in the air. Can they have neglected to accommodate themselves to the demands of their environment? If evolution and environment cause the topsy-turvy progress of the one, I have the right, if words have any meaning whatever, to demand as much of the others, since their organization is so much alike and their mode of life identical.

I have but little respect for theories which, when confronted with two similar cases, are unable to interpret the one without contradicting the other. They make me laugh when they become merely childish. For example: why has the tiger a coat streaked black and yellow? A matter of environment, replies one of our evolutionary masters. Ambushed in bamboo thickets where the golden radiance of the sun is intersected by stripes of shadow cast by the foliage, the animal, the better to conceal itself, assumed the colour of its environment. The rays of the sun produced the tawny yellow of the coat; the stripes of shadow added the black bars.

And there you have it. Any one who refuses to accept the explanation must be very hard to please. I am one of these difficult persons. If it were a dinner-table jest, made over the walnuts and the wine, I would willingly sing ditto; but alas and alack, it is uttered without a smile, in a solemn and magisterial manner, as the last word in science! Toussenel, in his day, asked the naturalists an insidious question. Why, he enquired, have Ducks a little curly feather on the rump? No one, so far as I know, had an answer for the teasing cross-examiner: evolution had not been invented then. In our time the reason why would be forthcoming in a moment, as lucid and as well-founded as the reason why of the tiger's coat.

Enough of childish nonsense. The Cetonia-grub walks on its back because it has always done so. The environment does not make the animal; it is the animal that is made for the environment. To this simple philosophy, which is quite antiquated nowadays, I will add another, which Socrates expressed in these words:

"What I know best is that I know nothing."

CHAPTER 6.

THE TACHYTES.

***

The family of Wasps whose name I inscribe at the head of this chapter has not hitherto, so far as I know, made much noise in the world. Its annals are limited to methodical classifications, which make very poor reading. The happy nations, men say, are those which have no history. I accept this, but I also admit that it is possible to have a history without ceasing to be happy. In the conviction that I shall not disturb its prosperity, I will try to substitute the living, moving insect for the insect impaled in a cork-bottomed box.

It has been adorned with a learned name, derived from the Greek Tachytes, meaning rapidity, suddenness, speed. The creature's godfather, as we see, had a smattering of Greek; its denomination is none the less unfortunate: intended to instruct us by means of a characteristic feature, the name leads us astray. Why is speed mentioned in this connection? Why a label which prepares the mind for an exceptional velocity and announces a race of peerless coursers? Nimble diggers of burrows and eager hunters the Tachytes are, to be sure, but they are no better than a host of rivals. Not the Sphex, nor the Ammophila, nor the Bembex, nor many another would admit herself beaten in either flying or running. At the nesting-season, all this tiny world of huntresses is filled with astounding activity. The quality of a speedy worker being common to all, none can boast of it to the exclusion of the rest.

Had I had a vote when the Tachytes was christened, I should have suggested a short, harmonious, well-sounding name, meaning nothing else than the thing meant. What better, for example, than the term Sphex? The ear is satisfied and the mind is not corrupted by a prejudice, a source of error to the beginner. I have not nearly as much liking for Ammophila, which represents as a lover of the sands an animal whose establishments call for compact soil. In short, if I had been forced, at all costs, to concoct a barbarous appellation out of Latin or Greek in order to recall the creature's leading characteristic, I should have attempted to say, a passionate lover of the Locust.

Love of the Locust, in the broader sense of the Orthopteron, an exclusive, intolerant love, handed down from mother to daughter with a fidelity which the centuries fail to impair, this, yes, this indeed depicts the Tachytes with greater accuracy than a name smacking of the racecourse. The Englishman has his roast-beef; the German his sauerkraut; the Russian his caviare; the Neapolitan his macaroni; the Piedmontese his polenta; the man of Carpentras his tian. The Tachytes has her Locust. Her national dish is also that of the Sphex, with whom I boldly associate her. The methodical classifier, who works in cemeteries and seems to fly the living cities, keeps the two families far removed from each other because of considerations and attaching to the nervures of the wings and the joints of the palpi. At the risk of passing for a heretic, I bring them together at the suggestion of the menu-card.

To my own knowledge, my part of the country possesses five species, one and all addicted to a diet of Orthoptera. Panzer's Tachytes (T. Panzeri, VAN DER LIND), girdled with red at the base of the abdomen, must be pretty rare. I surprise her from time to time working on the hard roadside banks and the trodden edges of the footpaths. There, to a depth of an inch at most, she digs her burrows, each isolated from the rest. Her prey is an adult, medium-sized Acridian, such as the White-banded Sphex pursues. The captive of the one would not be despised by the other. Gripped by the antennae, according to the ritual of the Sphex, the victim is trailed along on foot and laid beside the nest, with the head pointing towards the opening. The pit, prepared in advance, is closed for the time being with a tiny flagstone and some bits of gravel, in order to avoid either the invasion of a passerby or obstruction by landslips during the huntress' absence. A like precaution is taken by the White-banded Sphex. Both observe the same diet and the same customs.

The Tachytes clears the entrance to the home and goes in alone. She returns, puts out her head and, seizing her prey by the antennae, warehouses it by dragging backwards. I have repeated, at her expense, the tricks which I used to play on the Sphex. While the Tachytes is underground, I move the game away. The insect comes up again and sees nothing at its door; it comes out and goes to fetch its Locust, whom it places in position as before. This done, it goes in again by itself. In its absence I once more pull back the prey. Fresh emergence of the Wasp, who puts things to rights and persists in going down again, still by herself, however often I repeat the experiment. Yet it would be very easy for her to put an end to my teasing: she would only have to descend straightway with her game, instead of leaving it for a moment on her doorstep. But, faithful to the usages of her race, she behaves as her ancestors behaved before her, even though the ancient custom happen to be unprofitable. Like the Yellow-winged Sphex, whom I have teased so often during her cellaring-operations, she is a narrow conservative, learning nothing and forgetting nothing.

Let us leave her to do her work in peace. The Locust disappears underground and the egg is laid upon the breast of the paralysed insect. That is all: one carcase for each cell, no more. The entrance is stopped at last, first with stones, which will prevent the trickling of the embankment into the chamber; next with sweepings of dust, under which every vestige of the subterranean house disappears. It is now done: the Tachytes will come here no more. Other burrows will occupy her, distributed at the whim of her vagabond humour.

A cell provisioned before my eyes on the 22nd of August, in one of the walls in the harmas, contained the finished cocoon a week later. I have not noted many examples of so rapid a development. This cocoon recalls, in its shape and texture, that of the Bembex-wasps. It is hard and mineralized, this is to say, the warp and woof of silk are hidden by a thick encrustation of sand. This composite structure seems to me characteristic of the family; at all events I find it in the three species whose cocoons I know. If the Tachytes are nearly related to the Spheges in diet, they are far removed from them in the industry of their larvae. The first are workers in mosaic, encrusting a network of silk and sand; the second weave pure silk.

Of smaller size and clad in black with trimmings of silvery down on the edge of the abdominal segments, the Tarsal Tachytes frequents the ledges of soft limestone in fairly populous colonies. (T. tarsina, LEP.)[1] August and September are the season of her labours. Her burrows, very close to one another when an easily-worked vein presents itself, afford an ample harvest of cocoons once the site is discovered. In a certain gravel-pit in the neighbourhood, with vertical walls visited by the sun, I have been able within a short space of time to collect enough to fill the hollow of my hand completely. They differ from the cocoons of the preceding species only in their smaller size. The provisions consist of young Acridians, varying from about a quarter to half an inch in length. The adult insect does not appear in the assorted bags of game, being no doubt too tough for the feeble grub. All the carcases consist of Locust-larvae, whose budding wings leave the back uncovered and put one in mind of the short skirts of a skimpy jacket. Small so that it may be tender, the game is numerous so that it may suffice all needs. I count from two to four carcases to a cell. When the time comes we will discover the reason for these differences in the rations served.

---

1 According to M. J. Perez, to whom I submitted the Wasp of which I am about to speak, this Tachytes might well be a new species, if it is not Lepelletier's T. tarsina or its equivalent, Panzer's T. unicolor. Any one wishing to clear up this point will always recognize the quarrelsome insect by its behaviour. A minute description seems useless to me in the type of investigation which I am pursuing.

The Mantis-killing Tachytes wears a red scarf, like her kinswoman, Panzer's Tachytes.[2] I do not think that she is very widely distributed. I made her acquaintance in the Serignan woods, where she inhabits, or rather used to inhabit – for I fear that I have depopulated and even destroyed the community by my repeated excavations – where she used to inhabit one of those little mounds of sand which the wind heaps up against the rosemary clumps. Outside this small community, I never saw her again. Her history, rich in incident, will be given with all the detail which it deserves. I will confine myself for the moment to mentioning her rations, which consist of Mantis-larvae, those of the Praying Mantis predominating. My lists record from three to sixteen heads for each cell. Once again we note a great inequality of rations, the reason for which we must try to discover.

What shall I say of the Black Tachytes. I have there described her contests with the Sphex, whose burrow she seems to me to have usurped; I show her dragging along the ruts in the roads a paralysed Cricket, seized by the hauling-ropes, his antennae; I speak of her hesitations, which lead me to suspect her for a homeless vagabond, and finally on her surrender of her game, with

---

2 The Mantis-hunting Tachytes was submitted to examination by M. J. Perez, who failed to recognize her. This species may well be new to our fauna. I confine myself to calling her the Mantis-killing Tachytes and leave to the specialists the task of adorning her with a Latin name, if it be really the fact that the Wasp is not yet catalogued. I will be brief in my delineation. To my thinking the best description is this: mantis-hunter. With this information it is impossible to mistake the insect, in my district of course. I may add that it is black, with the first two abdominal segments, the legs and the tarsi a rusty red. Clad in the same livery and much smaller than the female, the male is remarkable for his eyes, which are of a beautiful lemon-yellow when he is alive. The length is nearly half an inch for the female and a little more than half this for the male.

which she seems at once satisfied and embarrassed. Save for the dispute with the Sphex, an unique event in my records as observer, I have seen all the rest many a time, but never anything more. The Black Tachytes, though the most frequent of all in my neighbourhood, remains a riddle to me. I know nothing of her dwelling, her larvae, her cocoons, her family-arrangements. All that I can affirm, judging by the invariable nature of the prey which one sees her dragging along, is that she must feed her larvae on the same non-adult Cricket that the Yellow-winged Sphex chooses for hers.

Is she a poacher, a pillager of other's property, or a genuine huntress? My suspicions are persistent, though I know how chary a man should be of suspicions. At one time I had my doubts about Panzer's Tachytes, whom I grudged a prey to which the White-banded Sphex might have laid claim. today I have no such doubts: she is an honest worker and her game is really the result of her hunting. While waiting for the truth to be revealed and my suspicions set aside, I will complete the little that I know of her by noting that the Black Tachytes passes the winter in the adult form and away from her cell. She hibernates, like the Hairy Ammophila. In warm, sheltered places, with low, perpendicular, bare banks, dear to the Wasps, I am certain of finding her at any time during the winter, however briefly I investigate the earthen surface, riddled with galleries. I find the Tachytes cowering singly in the hot oven formed by the end of a tunnel. If the temperature be mild and the sky clear, she emerges from her retreat in January and February and comes to the surface of the bank to see whether spring is making progress. When the shadows fall and the heat decreases, she reenters her winter-quarters.

The Anathema Tachytes (T. anathema, VAN DER LIND), the giant of her race, almost as large as the Languedocian Sphex and, like her, decorated with a red scarf round the base of the abdomen, is rarer than any of her congeners. I have come upon her only some four or five times, as an isolated individual and always in circumstances which will tell us of the nature of her game with a probability that comes very near to certainty. She hunts underground, like the Scoliae. In September I see her go down into the soil, which has been loosened by a recent light shower; the movement of the earth turned over keeps me informed of her subterranean progress. She is like the Mole, ploughing through a meadow in pursuit of his White Worm. She comes out farther on, nearly a yard from the spot at which she went in. This long journey underground has taken her only a few minutes.

Is this due to extraordinary powers of excavation on her part? By no means: the Anathema Tachytes is an energetic tunneller, no doubt, but, after all, is incapable of performing so great a labour in so short a time. If the underground worker is so swift in her progress, it is because the track followed has already been covered by another. The trail is ready prepared. We will describe it, for it is clearly defined before the intervention of the Wasp.

On the surface of the ground, for a length of two paces at most, runs a sinuous line, a beading of crumbled soil, roughly the width of my finger. From this line of ramifications (others) shoot out to left and right, much shorter and irregularly distributed. One need not be a great entomological scholar to recognize, at the first glance, in these pads of raised earth, the trail of a

Mole-cricket, the Mole among insects. It is the Mole-cricket who, seeking for a root to suit her, has excavated the winding tunnel, with investigation-galleries grafted to either side of the main road. The passage is free therefore, or at most blocked by a few landslips, of which the Tachytes will easily dispose. This explains her rapid journey underground.

But what does she do there? For she is always there, in the few observations which chance affords me. A subterranean excursion would not attract the Wasp if it had no object. And its object is certainly the search for some sort of game for her larvae. The inference becomes inevitable: the Anathema Tachytes, who explores the Mole-cricket's galleries, gives her larvae this same Mole-cricket as their food. Very probably the specimen selected is a young one, for the adult insect would be too big. Besides, to this consideration of quantity is added that of quality. Young and tender flesh is highly appreciated, as witness the Tarsal Tachytes, the Black Tachytes and the Mantis-killing Tachytes, who all three select game that is not yet made tough by age. It goes without saying that the moment the huntress emerged from the ground I proceeded to dig up the track. The Mole-cricket was no longer there. The Tachytes had come too late; and so had I.

Well, how right was I to define the Tachytes as a Locust lover! What constancy in the gastronomic rules of the race! And what tact in varying the game, while keeping within the order of the Orthoptera! What have the Locust, the Cricket, the Praying Mantis and the Mole-cricket in common, as regards their general appearance? Why, absolutely nothing! None of us, if he were unfamiliar with the delicate associations dictated by anatomy, would think of classing them together. The Tachytes, on the other hand, makes no mistake. Guided by her instinct, which rivals the science of a Latreille, she groups them all together.

This instinctive taxonomy becomes more surprising still if we consider the variety of the game stored in a single burrow. The Mantis-killing Tachytes, for instance, preys indiscriminately upon all the Mantides that occur in her neighbourhood. I see her warehousing three of them, the only varieties, in fact, that I know in my district. They are the following: the Praying Mantis and the Empusa. The numerical predominance in the Tachytes' cells belongs to the Praying Mantis; and the Grey Mantis occupies second place. The Empusa, who is comparatively rare on the brushwood in the neighbourhood, is also rare in the storehouses of the Wasp; nevertheless her presence is repeated often enough to show that the huntress appreciates the value of this prey when she comes across it. The three sorts of game are in the larval state, with rudimentary wings. Their dimensions, which vary a good deal, fluctuate between two-fifths and four-fifths of an inch in length.

The Praying Mantis is a bright green; she boasts an elongated prothorax and an alert gait. The other Mantis is ash-grey. Her prothorax is short and her movements heavy. The coloration therefore is no guide to the huntress, any more than the gait. The green and the grey, the swift and the slow are unable to baffle her perspicacity. To her, despite the great difference in appearance, the two victims are Mantes. And she is right.

But what are we to say of the Empusa? The insect world, at all events in our parts, contains no more fantastic creature. The children here, who are remarkable for finding names which really depict the animal, call the larva "the Devilkin." It is indeed a spectre, a diabolical phantom worthy of the pencil of a Callot. There is nothing to beat it in the extravagant medley of figures in his "Temptation of Saint Anthony." Its flat abdomen, scalloped at the edges, rises into a twisted crook; its peaked head carries on the top two large, divergent, tusk-shaped horns; its sharp, pointed face, which can turn and look to either side, would fit the wily purpose of some Mephistopheles; its long legs have cleaver-like appendages at the joints, similar to the arm-pieces which the knights of old used to bear upon their elbows. Perched high upon the shanks of its four hind-legs, with its abdomen curled, its thorax raised erect, its front-legs, the traps and implements of warfare, folded against its chest, it sways limply from side to side, on the tip of the bough.

Any one seeing it for the first time in its grotesque pose will give a start of surprise. The Tachytes knows no such alarm. If she catches sight of it, she seizes it by the neck and stabs it. It will be a treat for her children. How does she manage to recognize in this spectre the near relation of the Praying Mantis? When frequent hunting-expeditions have familiarized her with the last-named and suddenly, in the midst of the chase, she encounters the Devilkin, how does she become aware that this strange find makes yet another excellent addition to her larder? This question, I fear, will never receive an adequate reply. Other huntresses have already set us the problem; others will set it to us again. I shall return to it, not to solve it, but to show even more plainly how obscure and profound it is. But we will first complete the story of the Mantis-killing Tachytes.

The colony which forms the subject of my investigations is established in a mound of fine sand which I myself cut into, a couple of years ago, in order to unearth a few Bembex larvae. The entrances to the Tachytes' dwelling open upon the little upright bank of the section. At the beginning of July the work is in full swing. It must have been going on already for a week or two, for I find very forward larvae, as well as recent cocoons. There are here, digging into the sand or returning from expeditions with their booty, some hundred females, whose burrows, all very close to one another, cover an area of barely a square yard. This hamlet, small in extent, but nevertheless densely populated, shows us the Mantis-slayer under a moral aspect which is not shared by the Locust slayer, Panzer's Tachytes, who resembles her so closely in costume. Though engaged in individual tasks, the first seeks the society of her kind, as do certain of the Sphex-wasps, while the second establishes herself in solitude, after the fashion of the Ammophila. Neither the personal form nor the nature of the occupation determines sociability.

Crouching voluptuously in the sun, on the sand at the foot of the bank, the males lie waiting for the females, to plague them as they pass. They are ardent lovers, but cut a poor figure. Their linear dimensions are barely half those of the other sex, which implies a volume only one-eighth as great. At a short distance they appear to wear on their heads a sort of gaudy turban. At close quarters this headgear is seen to consist of the eyes, which are very large and a bright lemon-yellow and which almost entirely surround the head.

At ten o'clock in the morning, when the heat begins to grow intolerable to the observer, there is a continual coming and going between the burrows and the tufts of grass, everlasting, thyme and wormwood, which constitute the Tachytes' hunting-grounds within a moderate radius. The journey is so short that the Wasp brings her game home on the wing, usually in a single flight. She holds it by the forepart, a very judicious precaution, which is favourable to rapid stowage in the warehouse, for then the Mantis' legs stretch backwards, along the axis of the body, instead of folding and projecting sideways, when their resistance would be difficult to overcome in a narrow gallery. The lanky prey dangles beneath the huntress, all limp, lifeless and paralysed. The Tachytes, still flying, alights on the threshold of the home and immediately, contrary to the custom of Panzer's Tachytes, enters with her prey trailing behind her. It is not unusual for a male to come upon the scene at the moment of the mother's arrival. He is promptly snubbed. This is the time for work, not for amusement. The rebuffed male resumes his post as a watcher in the sun; and the housewife stows her provisions.

But she does not always do so without hindrance. Let me recount one of the misadventures of this work of storage. There is in the neighbourhood of the burrows a plant which catches insects with glue. It is the Oporto silene (S. portensis), a curious growth, a lover of the seaside dunes, which, though of Portuguese origin, as its name would seem to indicate, ventures inland, even as far as my part of the country, where it represents perhaps a survivor of the coastal flora of what was once a Pliocene sea. The sea has disappeared; a few plants of its shores have remained behind. This Silene carries in most of its internodes, in those both of the branches and of the main stalk, a viscous ring, two to four-fifths of an inch wide, sharply delimited above and below. The coating of glue is of a pale brown. Its stickiness is so great that the least touch is enough to hold the object. I find Midges, Plant-lice and Ants caught in it, as well as tufted seeds which have blown from the capitula of the Cichoriaceae. A gadfly, as big as a Blue bottle, falls into the trap before my eyes. She has barely alighted on the perilous perch when lo, she is held by the hinder tarsi! The Fly makes violent efforts to take wing; she shakes the slender plant from top to bottom. If she frees her hinder tarsi she remains snared by the front tarsi and has to begin all over again. I was doubting the possibility of her escape when, after a good quarter of an hour's struggle, she succeeded in extricating herself.

But, where the gadfly has got off, the Midge remains. The winged Aphis also remains, the Ant, the Mosquito and many another of the smaller insects. What does the plant do with its captures? Of what use are these trophies of corpses hanging by a leg or a wing? Does the vegetable bird-limer, with its sticky rings, derive advantage from these death-struggles? A Darwinian, remembering the carnivorous plants, would say yes. As for me, I don't believe a word of it. The Oporto silene is ringed with bands of gum. Why? I don't know. Insects are caught in these snares. Of what use are they to the plant? Why, none at all; and that's all about it. I leave to others, bolder than myself, the fantastic idea of taking these annular exudations for a digestive fluid which will reduce the captured Midges to soup and make them serve to feed the Silene. Only I warn them that the insects sticking to the plant do not dissolve into broth, but shrivel, quite uselessly, in the sun.

Let us return to the Tachytes, who is also a victim of the vegetable snare. With a sudden flight, a huntress arrives, carrying her drooping prey. She grazes the Silene's lime-twigs too closely. Behold the Mantis caught by the abdomen. For twenty minutes at least the Wasp, still on the wing, tugs at her, tugging again and again, to overcome the cause of the hitch and release the spoil. The hauling-method, a continuation of the flight, comes to nothing; and no other is attempted. At last the insect wearies and leaves the Mantis hanging to the Silene.

Now or never was the moment for the intervention of that tiny glimmer of reason which Darwin so generously grants to animals. Do not, if you please, confound reason with intelligence, as people are too prone to do. I deny the one; and the other is incontestable, within very modest limits. It was, I said, the moment to reason a little, to discover the cause of the hitch and to attack the difficulty at its source. For the Tachytes the matter was of the simplest. She had but to grab the body by the skin of the abdomen immediately above the spot caught by the glue and to pull it towards her, instead of persevering in her flight without releasing the neck. Simple though this mechanical problem was, the insect was unable to solve it, because she was not able to trace the effect back to the cause, because she did not even suspect that the stoppage had a cause.

Ants doting on sugar and accustomed to cross a footbridge in order to reach the warehouse are absolutely prevented from doing so when the bridge is interrupted by a slight gap. They would only need a few grains of sand to fill the void and restore the causeway. They do not for a moment dream of it, plucky navvies though they be, capable of raising miniature mountains of excavated soil. We can get them to give us an enormous cone of earth, an instinctive piece of work, but we shall never obtain the juxtaposition of three grains of sand, a reasoned piece of work. The Ant does not reason, any more than the Tachytes.

If you bring up a tame Fox and set his platter of food before him, this creature of a thousand tricks confines himself to tugging with all his might at the leash which keeps him a step or two from his dinner. He pulls as the Tachytes pulls, exhausts himself in futile efforts and then lies down, with his little eyes leering fixedly at the dish. Why does he not turn round? This would increase his radius; and he could reach then the food with his hind-foot and pull it towards him. The idea never occurs to him. Yet another animal deprived of reason.

Friend Bull, my Dog, is no better-endowed, despite his quality as a candidate for humanity. In our excursions through the woods, he happens to get caught by the paw in a wire snare set for rabbits. Like the Tachytes, he tugs at it obstinately and only pulls the noose tighter. I have to release him when he does not himself succeed in snapping the wire by his hard pulling. When he tries to leave the room, if the two leaves of the door are just ajar, he contents himself with pushing his muzzle, like a wedge, into the too narrow aperture. He moves forward, pushing in the direction which he wishes to take. His simple, doglike method has one unfailing result: the two leaves of the door, when pushed, merely shut still closer. It would be easy for him to pull one of them towards him with his paw, which would make the passage wider; but this would be a movement backward, contrary to his natural impulse; and so he does not think of it. Yet another creature that does not reason.

The Tachytes, who stubbornly persists in tugging at her limed Mantis and refuses to acknowledge any other method of wresting her from the Silene's snare, shows us the Wasp in an unflattering light. What a very poor intellect! The insect becomes only the more wonderful, therefore, when we consider its supreme talent as an anatomist. Many a time I have insisted upon the incomprehensible wisdom of instinct; I do so again at the risk of repeating myself. An idea is like a nail: it is not to be driven in save by repeated blows. By hitting it again and again, I hope to make it enter the most rebellious brains. This time I shall attack the problem from the other end, that is, I shall first allow human knowledge to have its say and shall then interrogate the insect's knowledge.

The outward structure of the Praying Mantis would of itself be enough to teach us the arrangement of the nerve-centres which the Tachytes has to injure in order to paralyse its victim, which is destined to be devoured alive but harmless. A narrow and very long prothorax divides the front pair of legs from the two hinder pairs. There must therefore be an isolated ganglion in front and two ganglia, close to each other, about two-fifths of an inch back. Dissection confirms this forecast completely. It shows us three fairly bulky thoracic ganglia, arranged in the same manner as the legs. The first which actuates the forelegs, is placed opposite their roots. It is the largest of the three. It is also the most important, for it presides over the insect's weapons, over the two powerful arms, toothed like saws and ending in harpoons. The other two, divided from the first by the whole length of the prothorax, each face the origin of the corresponding legs; consequently they are very near each other. Beyond them are the abdominal ganglia, which I pass over in silence, as the operating insect does not have to trouble about them. The movements of the belly are mere pulsations and are in no way dangerous.

Now let us do a little reasoning on behalf of our non-reasoning insect. The sacrificer is weak; the victim is comparatively powerful. Three strokes of the lancet must abolish all offensive movement. Where will the first stroke be delivered? In front is a real engine of warfare, a pair of powerful shears with toothed jaws. Let the forearm close upon the upper arm; and the imprudent insect, crushed between the two saw-blades, will be torn to pieces; wounded by the terminal hook, it will be eviscerated. This ferocious mechanism is the great danger; it is this that must be mastered at the outset, at the risk of life; the rest is less urgent. The first blow of the stylet, cautiously directed, is therefore aimed at the lethal forelegs, which imperil the vivisector's own existence. Above all, there must be no hesitation. The blow must be accurate then and there, or the sacrificer will be caught in the vice and perish. The two other pairs of legs present no danger to the operator, who might neglect them if she had only her own security to think of; but the surgeon is operating with a view to the egg, which demands complete immobility in the provisions. Their centres of innervation will therefore be stabbed as well, with the leisure which the Mantis, now put out of action, permits. These legs, as well as their nervous centres, are situated very far behind the first point attacked. There is a long neutral interval, that of the prothorax, into which it is quite useless to drive the sting. This interval has to be crossed; by a backward movement conforming with the secrets of the victim's internal anatomy, the second ganglion must be reached and then its neighbour, the third. In short, the surgical operation may be formulated thus: a first stab of the lancet in front; a considerable movement to the rear, measuring about two-fifths of an inch; lastly, two lancet-thrusts at two points very close together. Thus speaks the science of man; thus counsels reason, guided by anatomical structure. Having said this much let us observe the insect's practice.

There is no difficulty about seeing the Tachytes operate in our presence; we have only to resort to the method of substitution, which has already done me so much service, that is, to deprive the huntress of her prey and at once to give her, in exchange, a living Mantis of about the same size. This substitution is impracticable with the majority of the Tachytes, who reach the threshold of their dwelling in a single flight and at once vanish underground with their game. A few of them, from time to time, harassed perhaps by their burden, chance to alight at a short distance from their burrow, or even drop their prey. I profit by these rare occasions to witness the tragedy.

The dispossessed Wasp recognizes instantly, from the proud bearing of the substituted Mantis, that she is no longer embracing and carrying off an inoffensive carcase. Her hovering, hitherto silent, develops a buzz, perhaps to overawe the victim; her flight becomes an extremely rapid oscillation, always behind the quarry. It is as who should say the quick movement of a pendulum swinging without a wire to hang from. The Mantis, however, lifts herself boldly upon her four hind-legs; she raises the forepart of her body, opens, closes and again opens her shears and presents them threateningly at the enemy; using a privilege which no other insect shares, she turns her head this way and that, as we do when we look over our shoulders; she faces her assailant, ready to strike a return blow wheresoever the attack may come. It is the first time that I have witnessed such defensive daring. What will be the outcome of it all?

The Wasp continues to oscillate behind the Mantis, in order to avoid the formidable grappling-engine; then, suddenly, when she judges that the other is baffled by the rapidity of her manoeuvres, she hurls herself upon the insect's back, seizes its neck with her mandibles, winds her legs round its thorax and hastily delivers a first thrust of the sting, to the front, at the root of the lethal legs. Complete success! The deadly shears fall powerless. The operator then lets herself slip as she might slide down a pole, retreats along the Mantis' back and, going a trifle lower, less than a finger's breadth, she stops and paralyses, this time without hurrying herself, the two pairs of hind-legs. It is done: the patient lies motionless; only the tarsi quiver, twitching in their last convulsions. The sacrificer brushes her wings for a moment and polishes her antennae by passing them through her mouth, an habitual sign of tranquillity returning after the emotions of the conflict; she seizes the game by the neck, takes it in her legs and flies away with it.

What do you say to it all? Do not the scientist's theory and the insect's practice agree most admirably? Has not the animal accomplished to perfection what anatomy and physiology enabled us to foretell? Instinct, a gratuitous attribute, an unconscious inspiration, rivals knowledge, that most costly acquisition. What strikes me most is the sudden recoil after the first thrust of the sting. The Hairy Ammophila, operating on her caterpillar, likewise recoils, but progressively, from one segment to the next.

Her deliberate surgery might receive a quasi-explanation if we ascribe it to a certain uniformity. With the Tachytes and the Mantis this paltry argument escapes us. Here are no lancet-pricks regularly distributed; on the contrary, the operating-method betrays a lack of symmetry which would be inconceivable, if the organization of the patient did not serve as a guide. The Tachytes therefore knows where her prey's nerve-centres lie; or, to speak more correctly, she behaves as though she knew.

This science which is unconscious of itself has not been acquired, by her and by her race, through experiments perfected from age to age and habits transmitted from one generation to the next. It is impossible, I am prepared to declare a hundred times, a thousand times over, it is absolutely impossible to experiment and to learn an art when you are lost if you do not succeed at the first attempt. Don't talk to me of atavism, of small successes increasing by inheritance, when the novice, if he misdirected his weapon, would be crushed in the trap of the two saws and fall a prey to the savage Mantis! The peaceable Locust, if missed, protests against the attack with a few kicks; the carnivorous Mantis, who is in the habit of feasting on Wasps far more powerful than the Tachytes, would protest by eating the bungler; the game would devour the hunter, an excellent catch. Mantis-paralysing is a most perilous trade and admits of no half-successes; you have to excel in it from the first, under pain of death. No, the surgical art of the Tachytes is not an acquired art. Whence then does it come, if not from the universal knowledge in which all things move and have their being!

What would happen if, in exchange for her Praying Mantis, I were to give the Tachytes a young Grasshopper? In rearing insects at home, I have already noted that the larvae put up very well with this diet; and I am surprised that the mother does not follow the example of the Tarsal Tachytes and provide her family with a skewerful of Locusts instead of the risky prey which she selects. The diet would be practically the same; and the terrible shears would no longer be a danger. With such a patient would her operating-method remain the same; should we again see a sudden recoil after the first stab under the neck; or would the vivisector modify her art in conformity with the unfamiliar nervous organization?

This second alternative is highly improbable. It would be nonsense to expect to see the paralyser vary the number and the distribution of the wounds according to the genus of the victim. Supremely skilled in the task that has fallen to its lot, the insect knows nothing further.

The first alternative seems to offer a certain chance and deserves a test. I offer the Tachytes, deprived of her Mantis, a small Grasshopper, whose hind-legs I amputate to prevent his leaping. The disabled Acridian jogs along the sand. The Wasp flies round him for a moment, casts a contemptuous glance upon the cripple and withdraws without attempting action. Let the prey offered be large or small, green or grey, short or long, rather like the Mantis or quite different, all my efforts miscarry. The Tachytes recognizes in an instant that this is no business of hers; this is not her family game; she goes off without even honouring my Grasshoppers with a peck of her mandibles.

This stubborn refusal is not due to gastronomical causes. I have stated that the larvae reared by my own hands feed on young Grasshoppers as readily as on young Mantes; they do not seem to perceive any difference between the two dishes; they thrive equally on the game chosen by me and that selected by the mother. If the mother sets no value on the Grasshopper, what then can be the reason of her refusal? I can see only one: this quarry, which is not hers, perhaps inspires her with fear, as any unknown thing might do; the ferocious Mantis does not alarm her, but the peaceable Grasshopper terrifies her. And then, if she were to overcome her apprehensions, she does not know how to master the Acridian and, above all, how to operate upon him. To every man his trade, to every Wasp her own way of wielding her sting. Modify the conditions ever so slightly; and these skilful paralysers are at an utter loss.

To every insect also its own art of fashioning the cocoon, an art which varies greatly, an art in which the larva displays all the resources of its instincts. The Tachytes, the Bembeces, the Stizi, the Palari and other burrowers build composite cocoons, hard as fruit-stones, formed of an encrustation of sand in a network of silk. We are already acquainted with the work of the Bembex. I will recall the fact that their larva first weaves a conical, horizontal bag of pure white silk, with wide meshes, held in place by interlaced threads which fix it to the walls of the cell. I have compared this bag, because of its shape, with a fishtrap. Without leaving this hammock, stretching its neck through the orifice, the worker gathers from without a little heap of sand, which it stores inside its workshop. Then, selecting the grains one by one, it encrusts them all around itself in the fabric of the bag and cements them with the fluid from its spinnerets, which hardens at once. When this task is finished, the house has still to be closed, for it has been wide open all this time to permit of the renewal of the store of sand as the heap inside becomes exhausted. For this purpose a cap of silk is woven across the opening and finally encrusted with the materials which the larva has retained at its disposal.

The Tachytes builds in quite another fashion, although its work, once finished, does not differ from that of the Bembex. The larva surrounds itself, to begin with, about the middle of its body with a silken girdle which a number of threads, very irregularly distributed, hold in place and connect with the walls of the cell. Sand is collected, within reach of the worker, on this general scaffolding. Then begins the work of minor masonry, with grains of sand for rubble and the secretion of the spinnerets for cement. The first course is laid upon the fore-edge of the suspensory ring. When the circle is completed, a second course of grains of sand, stuck together by the fluid silk, is raised upon the hardened edge of what has just been done. Thus the work proceeds, by ring-shaped courses, laid edge to edge, until the cocoon, having acquired half of its proper length, is rounded into a cap and finally is closed. The building-methods of the Tachytes-larva remind me of a mason constructing a round chimney, a narrow tower of which he occupies the centre. Turning on his own axis and using the materials placed to his hand, he encloses himself little by little in his sheath of masonry. In the same way the worker encloses itself in its mosaic. To build the second half of the cocoon, the larva turns round and builds in the same way on the other edge of the original ring. In about thirty-six hours the solid shell is completed.

I am rather interested to see the Bembex and the Tachytes, two workers in the same guild, employ such different methods to achieve the same result. The first begins by weaving an eel-trap of pure silk and next encrusts the grains of sand inside; the second, a bolder architect, is economical of the silk envelope, confines itself to a hanging girdle and builds course by course. The building-materials are the same: sand and silk; the surroundings amid which the two artisans work are the same: a cell in a soil of sandy gravel; yet each of the builders possesses its individual art, its own plan, its one method.

The nature of the food has no more effect upon the larva's talents than the environment in which it lives or the materials employed. The proof of this is furnished by Stiza ruficornis, another builder of cocoons in grains of sand cemented with silk. This sturdy Wasp digs her burrows in soft sandstone. Like the Mantis-killing Tachytes, she hunts the various Mantides of the countryside, consisting mainly of the Praying Mantis; only her large size requires them to be more fully developed, without however having attained the form and the dimensions of the adult. She places three to five of them in each cell.

In solidity and volume her cocoon rivals that of the largest Bembex; but it differs from it, at first sight, by a singular feature of which I know no other example. From the side of the shell, which is uniformly smoothed on every side, a rough knob protrudes, a little clod of sand stuck on to the rest. The work of Stizus ruficornis can at once be recognized, among all the other cocoons of a similar nature, by this protuberance.

Its origin will be explained by the method which the larva follows in constructing its strongbox. At the beginning, a conical bag is woven of pure white silk; you might take it for the initial eel-trap of the Bembeces, only this bag has two openings, a very wide one in front and another, very narrow one at the side. Through the front opening the Stizus provides itself with sand as and when it spends this material on encrusting the interior. This strengthens the cocoon; and the cap which closes it is made next. So far it is exactly like the work of the Bembex. We now have the worker enclosed, engaged in perfecting the inner wall. For these final touches a little more sand is needed. It obtains it from outside by means of the aperture which it has taken the precaution of contriving in the side of its building, a narrow dormer-window just large enough to allow its slender neck to pass. When the store has been taken in, this accessory orifice, which is used only during the last few moments, is closed with a mouthful of mortar, thrust outward from within. This forms the irregular nipple which projects from the side of the shell.

For the present I shall not expatiate further upon Stizus ruficornis, whose complete biography would be out of place in this chapter. I will limit myself to mentioning its method of constructing strongboxes in order to compare it with that of the Bembex and above all with that of the Tachytes, a consumer, like itself, of Praying Mantes. From this parallel it seems to me to follow that the conditions of life in which men see today the origin of instincts – the type of food, the surroundings amid which the larval life is passed, the materials available for a defensive wrapper and other factors which the evolutionists are accustomed to invoke –

have no actual influence upon the larva's industry. My three architects in glued sand, even when all the conditions, down to the nature of the provisions, are the same, adopt different means to execute an identical task. They are engineers who have not graduated from the same school, who have not been educated on the same principles, though the lesson of things is almost the same for all of them. The workshop, the work, the provisions have not determined the instinct. The instinct comes first; it lays down laws instead of being subject to them.

CHAPTER 7.

CHANGE OF DIET.

***

Brillat-Savarin, when pronouncing his famous maxim, "Tell me what you eat and I will tell you what you are," certainly never suspected the signal confirmation which the entomological world would bestow upon his saying. Our gastrosopher was speaking only of the culinary caprices of man rendered fastidious by the sweets of life; but he might, in a more serious department of thought, have given his formula a wider and more general bearing and applied it to the dishes which vary so greatly according to latitude, climate and customs; he might above all have taken into his reckoning the harsh realities suffered by the common people, when perhaps his ideal of moral worth would have been found in a platter of chickpeas oftener than in a pot of pate de foie gras. No matter: his aphorism, the mere whimsical sally of an epicure, becomes an imperious truth if we forget the luxury of the table and look into what is eaten by the little world which swarms around us.

To each its mess. The cabbage Pieris consumes the pungent leaves of the Cruciferae as the food of her infancy; the Silkworm disdains any foliage other than that of the mulberry-tree. The Spurge Hawk-moth requires the caustic milk-sap of the tithymals: the Corn-weevil the grain of wheat; the Pea-weevil, the seeds of the Leguminosae; the Balaninus the hazelnut, the chestnut, the acorn; the Brachycera the clove of garlic. Each has its diet, each its plant; and each plant has its customary guests. Their relations are so precise that in many cases one might determine the insect by the vegetable which supports it, or the vegetable by the insect.

If you know the lily, you may name as a Crioceris the tiny scarlet Scarabaeid that inhabits it and peoples its leaves with larvae which keep themselves cool beneath an overcoat of ordure. If you know the Crioceris, you may name as a lily the plant which she devastates. It will not perhaps be the common or white lily, but some other representative of the same family - Turk's cap lily, orange lily, scarlet Martagon, lancifoliate lily, tiger-spotted lily, golden lily - hailing from the Alps or the Pyrenees, or brought from China or Japan. Relying on the Crioceris, who is an expert judge of exotic as well as of native Liliaceae, you may name as a lily the plant with which you are unacquainted and trust the word of this singular botanical master. Whether the flower be red, yellow, ruddy-brown or sown with crimson spots, characteristics so unlike the immaculate whiteness of the familiar flower, do not hesitate, adopt the name dictated by the Beetle. Where man is liable to mistake the insect is never mistaken.

This insect botany, a cause of such grievous tribulations, has always impressed the worker in the fields, who for all that, is a very indifferent observer. The man who was the first to see his cabbage-plot devastated by caterpillars made the acquaintance of the Pieris. Science completed the process, in its desire to serve a useful purpose or merely to seek truth for truth's sake; and today the relations between the insect and the plant form a collection of records as important from the philosophical as from the practical, agricultural point of view. What is much less familiar to us, because it touches us less nearly, is the zoology of the insect, that is to say, the selection which it makes, to feed its larva, of this or that animal species, to the exclusion of others. The subject is so vast that a volume were not sufficient to exhaust it; besides, data are lacking in the vast majority of cases. It is reserved for a still very distant future to raise this point of biology to the level already reached by the question of vegetable diet. It will be enough if I contribute a few observations scattered through my writings or my notes.

What does the Wasp addicted to a predatory life eat, of course in the larval state? Now, to begin with, we see natural sections which adopt as their prey different species of one and the same order, in one and the same group. Thus the Ammophilae hunt exclusively the larvae of the night-flying Moths. This taste is shared by the Eumenes, a very different genus. The Spheges and Tachytes are addicted to Orthoptera; the Cerceres, apart from a few exceptions, are faithful to the Weevil; both the Philanthi and the Palari capture only Hymenoptera; the Pompili specialize in hunting the Spider; the Astata revels in the flavour of Bugs; the Bembeces want Flies and nothing else; the Scoliae enjoy the monopoly of the Lamellicorn-grubs; the Pelopaei favour the young Epeirae, the Stizi vary in opinion: of the two in my neighbourhood, one, S. ruficornis, fills her larder with Mantes and the other, S. tridentatus, fills it with Cicadellae; lastly, the Crabronidae. levy tribute upon the rabble of the Muscidae.

Already you see what a magnificent classification of these game-hunters might be made with a faithfully listed bill of fare. Natural groups stand out, characterized merely by the identity of their victuals. I trust that the methodical science of the future will take account of these gastronomic laws, to the great relief of the entomological novice, who is too often hampered by the snares of the mouthparts, the antennae and the nervures of the wings. I call for a classification in which the insect's aptitudes, its diet, its industry and its habits shall take precedence of the shape of a joint in its antennae. It will come; but when?

If from generalities we descend to details, we shall see that the very species may, in many instances, be determined from the nature of its victuals. The number of burrows of Philanthus apivorus which I have inspected since I have been rummaging the hot roadside embankments, to enquire into their population, would seem hyperbolical were I able to state the figures. They must amount, it seems to me, to thousands. Well, in this multitude of food-stores, whether recent or ancient, uncovered for a purpose or encountered by chance, I have not once, not as often as once, discovered other remains than those of the Hive-bee: the imperishable wings, still connected in pairs, the cranium and thorax enveloped in a violet shroud, the winding-sheet which time throws over these relics. today as when I was a beginner, ever so long ago; in the north as in the south of the country which I explored; in mountainous regions as on the plains, the Philanthus follows an unvarying diet: she must have the Hive-bee, always the Bee and never any other, however closely various other kinds of game resemble the Bee in quality. If, therefore, when exploring sunny banks, you find beneath the soil a small parcel of mutilated Bees, that will be enough to point to the existence of a local colony of Philanthus apivorus. She alone knows the recipe for making potted Bee-meat. The Crioceris was but now teaching us all about the lily family; and here the mildewed body of the Bee tells us of the Philanthus and her lair.

Similarly the female Ephippiger helps us to identify the Languedocian Sphex: her relics, the cymbals and the long sabre, are the unmistakable sign of the cocoon to which they adhere. The black Cricket, with his red-braided thighs, is the infallible label of the Yellow-winged Sphex; the larva of Oryctes nasicornis tells us of the Garden Scolia as certainly as the best description; the Cetonia-grub proclaims the Two-banded Scolia and the larva of the Anoxia announces the Interrupted Scolia.

After these exclusive ones, who disdain to vary their meals, let us mention the eclectics, who, in a group which is generally well-defined, are able to select among different kinds of game appropriate to their bulk. The Great Cerceris favours above all Cleonus ophthalmicus, one of the largest of our Weevils; but at need she accepts the other Cleoni, as well as the kindred genera, provided that the capture be of an imposing size. Cerceris arenaria extends her hunting-grounds farther afield: any Weevil of average dimensions is to her a welcome capture. The Buprestis-hunting Cerceris adopts all the Buprestes indiscriminately, so long as they are not beyond her strength. The Crowned Philanthus (P. coronatus, FAB.) fills her underground warehouses with Halicti chosen among the biggest. Much smaller than her kinswoman, Philanthus raptor, LEP., stores away Halicti chosen among the less large species. Any adult Acridian approaching an inch in length suits the White-banded Sphex. The various tidae of the neighbourhood are admitted to the larder of Stizus ruficornis and of the Mantis-hunting Tachytes on the sole condition of being young and tender. The largest of our Bembeces (B. rostrata, FAB., and B. bidentata, VAN DER LIND) are eager consumers of gadflies. With these chief dishes they associate relishes levied indifferently from the rest of the Fly clan. The Sandy Ammophila (A. sabulosa, VAN DER LIND) and the Hairy Ammophila (A. hirsuta, KIRB.) cram into each burrow a single but corpulent caterpillar, always of the Moth tribe and varying greatly in coloration, which denotes distinct species. The Silky Ammophila (A. holosericea, VAN DER LIND.) has a better assorted diet. She requires for each banqueter three or four items, which include the Measuring-worms, or Loopers, and the caterpillars of ordinary Moths, all of which are equally appreciated. The Brown-winged Solenius (S. fascipennis, LEP.), who elects to dwell in the soft dead wood of old willow-trees, has a marked preference for Virgil's Bee, Eristalis tenax, willingly adding, sometimes as a side-dish, sometimes as the principal game, Helophilus pendulus, whose costume is very different. On the faith of indistinguishable remains, we must no doubt enter a number of other Flies in her game-book. The Golden-mouthed Hornet (Crabro chrysostomus, LEP.) another burrower in old willow-trees, prefers the Syrphi, without distinction of species. The Wandering Solenius (S. vagus, LEP.), an inmate of the dry bramble-stems and of the dwarf-elder, lays under contribution for her larder the genera Syritta, Sphaerophoria, Sarcophaga, Syrphus, Melanophora, Paragus and apparently many others. The species which recurs most frequently in my notes is Syritta pipiens.

Without pursuing this tedious list any farther, we plainly perceive the general result. Each huntress has her characteristic tastes, so much so that, when we know the bill of fare, we can tell the genus and very often the species of the guest, thus proving the proud truth of the maxim, "Tell me what you eat and I will tell you what you are."

There are some which always need the same prey. The offspring of the Languedocian Sphex religiously consume the Ephippiger, that family dish so dear to their ancestors and no less dear to their descendants; no innovation in the ancient usages can tempt them. Others are better suited by variety, for reasons connected with flavour or with facility of supply; but then the selection of the game is kept within fixed limits. A natural group, a genus, a family, more rarely almost a whole order: this is the hunting-ground beyond which poaching is strictly forbidden. The law is absolute; and one and all scrupulously refrain from transgressing it.

In the place of the Praying Mantis, offer the Mantis-hunting Tachytes an equivalent in the shape of a Locust. She will scorn the morsel, though it would seem to be of excellent flavour, seeing that Panzer's Tachytes prefers it to any other form of game. Offer her a young Empusa, who differs so widely from the Mantis in shape and colour: she will accept without hesitation and operate before your eyes. Despite its fantastic appearance, the Devilkin is instantly recognized by the Tachytes as a Mantid and therefore as game falling within her scope.

In exchange for her Cleonus, give to the Great Cerceris a Buprestis, the delight of one of her near kinsfolk. She will have nothing to say to the sumptuous dish. Accept that! She, a Weevil-eater! Never in this world! Present her with a Cleonus of a different species, or any other large Weevil, of a sort which she has most probably never seen before, since it does not figure on the inventory of the provisions in her burrows. This time there is no show of disdain: the victim is seized and stabbed in the regulation manner and forthwith stored away.

Try to persuade the Hairy Ammophila that Spiders have a nutty flavour, as Lalande asserts; and you will see how coldly your hints are received. Try merely to convince her that the caterpillar of a Butterfly is as good to eat as the caterpillar of a Moth. You will not succeed. But, if you substitute for her underground larva, which I suppose to be grey, another underground larva striped with black, yellow, rusty-red or any other tint, this change of coloration will not prevent her from recognizing, in the substituted dish, a victim to her liking, an equivalent of her Grey Worm.

So with the rest, so far as I have been able to experiment with them. Each obstinately refuses what is alien to her hunting-preserves, each accepts whatever belongs to them, always provided that the game substituted is much the same in size and development as that whereof the owner has been deprived. Thus the Tarsal Tachytes, an appreciative epicure of tender flesh, would not consent to replace her pinch of young Acridian-grubs with the one big Locust that forms the food of Panzer's Tachytes; and the latter, in her turn, would never exchange her adult Acridian for the other's menu of small fry. The genus and the species are the same, but the age differs; and this is enough to decide the question of acceptance or refusal.

When its depredations cover a somewhat extensive group, how does the insect manage to recognize the genera, the species composing her allotted portion and to distinguish them from the rest with an assured vision which the inventory of her burrows proves never to be at fault? Is it the general appearance that guides her? No, for in some Bembex-burrows we shall find Sphaerophoriae, those slender, thong-like creatures, and Bombylii, looking like velvet pincushions; no again, for in the pits of the Silky Ammophila we shall see, side by side, the caterpillar of the ordinary shape and the Measuring-worm, a living pair of compasses which progresses by alternately opening out and closing; no, once more, for in the storerooms of Stizus ruficornis and the Mantis-hunting Tachytes we see stacked beside the Mantis the Empusa, her unrecognizable caricature.

Is it the colouring? Not at all. There is no lack of instances. What a variety of hues and metallic reflections, distributed in a host of different fashions, appear in the Buprestes that are hunted by the Cerceris celebrated by Leon Dufour. A painter's palette, containing crushed gold, bronze, ruby and amethyst, would find it difficult to rival these sumptuous colours. Nevertheless the Cerceris makes no mistake: all this nation of insects, so indifferently attired, represents to her, as to the entomologist, the nation of the Buprestes. The inventory of the Hornet's larder will include Diptera clad in grey or russet frieze; others are girdled with yellow, flecked with white, adorned with crimson lines; others are steel-blue, ebony black, or coppery green; and underneath this variety of dissimilar costumes we find the invariable Fly.

Let us take a concrete example. Ferrero's Cerceris (C. Ferreri, VAN DER LIND) consumes Weevils. Her burrows are usually lined with Phynotomi and Sitones both an indeterminate grey, and Otiorhynchi, black or tan-coloured. Now I have sometimes happened to unearth from her cells a collection of veritable jewels which, thanks to their bright metallic lustre, made a most striking contrast with the sombre Otiorhynchus. These were the Rhynchites (R. betuleti), who roll the vine-leaves into cigars. Equally magnificent, some of them were azure blue, others copper gilt, for the cigar-roller has a twofold colouring. How did the Cerceris manage to recognize in these jewels the Weevil, the near relative of the vulgar Phynotomus? Any such encounters probably found her lacking in expert knowledge; her race cannot have handed down to her other than very indeterminate propensities, for she does not appear to make frequent use of the Rhynchites, as is proved by my infrequent discovery of them amid the mass of my numerous excavations. For the first time, perhaps, passing through a vineyard, she saw the rich Beetle gleaming on a leaf; it was not for her a dish in current consumption, consecrated by the ancient usages of the family. It was something novel, exceptional, extraordinary. Well, this extraordinary creature is recognized with certainty as a Weevil and stored away as such. The glittering cuirass of the Rhynchites goes to take its place beside the grey cloak of the Phynotomus. No, it is not the colour that guides the choice.

Neither is it the shape. Cerceris arenaria hunts any medium-sized Weevil. I should be putting the reader's patience to too great a test if I attempted to give in this place a complete inventory of the specimens identified in her larder. I will mention only two, which my latest searches around my village have revealed. The Wasp goes hunting on the holm-oaks of the neighbouring hills the Pubescent Brachyderes (B. pubescens) and the Acorn-weevil (Balaninus glandium). What have these two Beetles in common as regards shape? I mean by shape not the structural details which the classifier examines through his magnifying-glass, not the delicate features which a Latreille would quote when drawing up a technical description, but the general picture, the general outline that impresses itself upon the vision even of an untrained eye and makes the man who knows nothing of science and above all the child, a most perspicacious observer, connect certain animals together.

In this respect, what have the Brachyderes and the Balaninus in common in the eyes of the townsman, the peasant, the child or the Cerceris? Absolutely nothing. The first has an almost cylindrical figure; the second, squat, short and thickset, is conical in front and elliptical, or rather shaped like the ace of hearts, behind. The first is black, strewn with cloudy, mouse-grey spots; the second is yellow ochre. The head of the first ends in a sort of snout; the head of the second tapers into a curved beak, slender as a horsehair and as long as the rest of the body. The Brachyderes has a massive proboscis, cut off short; the Balaninus seems to be smoking an insanely long cigarette-holder.

Who would think of connecting two creatures so unlike, of calling them by the same name? Outside the professional classifiers, no one would dare to. The Cerceris, more perspicacious, knows each of them for a Weevil, a quarry with a concentrated nervous system, lending itself to the surgical feat of her single stroke of the lancet. After obtaining an abundant booty at the cost of the blunt-mouthed insect, with which she sometimes stuffs her cellars to the exclusion of any other fare, according to the hazards of the chase, she now suddenly sees before her the creature with the extravagant proboscis. Accustomed to the first, will she fail to know the second? By no means: at the first glance she recognizes it as her own; and the cell already furnished with a few Brachyderes receives its complement of Balanini. If these two species are to seek, if the burrows are far from the holm-oaks, the Cerceris will attack Weevils displaying the greatest variety of genus, species, form and coloration, levying tribute indifferently on Sitones, Cneorhini, Geonemi, Otiorhynchi, Strophosomi and many others.

In vain do I rack my brains merely to guess at the signs upon which the huntress relies as a guide, without going outside one and the same group, in the midst of such a variety of game; above all by what characteristics she recognizes as a Weevil the strange Acorn Balaninus, the only one among her victims that wears a long pipe-stem. I leave to evolutionism, atavism and other transcendental "isms" the honour and also the risk of explaining what I humbly recognize as being too far beyond my grasp. Because the son of the bird-catcher who imitates the call of his victims has been fed on roast Robins, Linnets and Chaffinches, shall we hastily conclude that this education through the stomach will enable him later, without other initiation than that of the spit, to know his way about the ornithological groups and to avoid confusing them when his turn comes to set his limed twigs? Will the digesting of a ragout of little birds, however often repeated by him or his ascendants, suffice to make him a finished bird-catcher? The Cerceris has eaten Weevil; her ancestors have all eaten Weevil, religiously. If you see in this the reason that makes the Wasp a Weevil-expert endowed with a perspicacity unrivalled save by that of a professional entomologist, why should you refuse to admit that the same consequences would follow in the bird-catcher's family?

I hasten to abandon these insoluble problems in order to attack the question of provisions from another point of view. Every Hunting Wasp is confined to a certain genus of game, which is usually strictly limited. She pursues her appointed quarry and regards anything outside it with suspicion and distaste. The tricks of the experimenter, who drags her prey from under her and flings her another in exchange, the emotions of the possessor deprived of her property and immediately recovering it, but under another form, are powerless to put her on the wrong scent. Obstinately she refuses whatever is alien to her portion; instantly she accepts whatever forms part of it. Whence arises this insuperable repugnance for provisions to which thc family is unaccustomed? Here we may appeal to experiment. Let us do so: its dictum is the only one that can be trusted.

The first idea that presents itself and the only one, I think, that can present itself is that the larva, the carnivorous nurseling, has its preferences, or we had better say its exclusive tastes. This kind of game suits it; that does not; and the mother provides it with food in conformity with its appetites, which are unchangeable in each species. Here the family dish is the gadfly; elsewhere it is the Weevil; elsewhere again it is the Cricket, the Locust and the Praying Mantis. Good in themselves, in a general way, these several victuals may be noxious to a consumer who is not used to them. The larva which dotes on Locust may find caterpillar a detestable fare; and that which revels in caterpillar may hold Locust in horror. It would be hard for us to discover in what manner Cricket-flesh and Ephippiger-flesh differ as juicy, nourishing foodstuffs; but it does not follow that the two Sphex-wasps addicted to this diet have not very decided opinions on the matter, or that each of them is not filled with the highest esteem for its traditional dish and a profound dislike for the other. There is no discussing tastes.

Moreover, the question of health may well be involved. There is nothing to tell us that the Spider, that treat for the Pompilus, is not poison, or at least unwholesome food, to the Bembex, the lover of gadflies; that the Ammophila's succulent caterpillar is not repugnant to the stomach of the Sphex fed upon the dry Acridian. The mother's esteem for one kind of game and her distrust of another would in that case be due to the likes and dislikes of her larvae; the victualler would regulate the bill of fare by the gastronomic demands of the victualled.

This exclusiveness of the carnivorous larva seems all the more probable inasmuch as the larva reared on vegetable food refuses in any way to lend itself to a change of diet. However pressed by hunger, the caterpillar of the Spurge Hawk-moth, which browses on the tithymals, will allow itself to starve in front of a cabbage leaf which makes a peerless meal for the Pieris. Its stomach, burned by pungent spices, will find the Crucifera insipid and uneatable, though its piquancy is enhanced by essence of sulphur. The Pieris, on its part, takes good care not to touch the tithymals: they would endanger its life. The caterpillar of the Death's-head Hawk-moth requires the solanaceous narcotics, principally the potato, and will have nothing else. All that is not seasoned with solanin it abhors. And it is not only larvae whose food is strongly spiced with alkaloids and other poisonous substances that refuse any innovation in their food; the others, even those whose diet is least juicy, are invincibly uncompromising. Each has its plant or its group of plants, beyond which nothing is acceptable.

I remember a late frost which had nipped the buds of the mulberry-trees during the night, just when the first leaves were out. Next day there was great excitement among my neighbours: the silkworms had hatched and the food had suddenly failed. The farmers had to wait for the sun to repair the disaster; but how were they to keep the famishing newborn grubs alive for a few days? They knew me for an expert in plants; by collecting them as I walked through the fields I had earned the name of a medical herbalist. With poppy-flowers I prepared an elixir which cleared the sight; with borage I obtained a syrup which was a sovran remedy for whooping-cough; I distilled camomile; I extracted the essential oil from the wintergreen. In short, botany had won for me the reputation of a quack doctor. After all, that was something.

The housewives came in search of me from every point of the compass and with tears in their eyes explained the situation. What could they give their silkworms while waiting for the mulberry to sprout afresh? It was a serious matter, well worthy of commiseration. One was counting on her batch to buy a length of cloth for her daughter, who was on the point of getting married; another told me of her plans for a Pig to be fattened against the coming winter; all deplored the handful of crown-pieces which, hoarded in the hiding-place in the cupboard, would have afforded help in difficult times.

And, full of their troubles, they unfolded, before my eyes, a scrap of flannel on which the vermin were swarming:

"Regardas, moussu! Venoun d'espeli; et ren per lour douna! Ah, pecaire!" "Look, sir! The frost has come and we've nothing to give them! Oh, what a misfortune!"

Poor people! What a harsh trade is yours: respectable above all others, but of all the most uncertain! You work yourselves to death; and, when you have almost reached your goal, a few hours of a cold night, which comes upon you suddenly, destroys your harvest. To help these afflicted ones seemed to me a very difficult thing. I tried, however, taking botany as my guide; it suggested to me, as substitutes for the mulberry, the members of closely-related families: the elm, the nettle-tree, the nettle, the pellitory. Their nascent leaves, chopped small, were offered to the silkworms. Other and far less logical attempts were made, in accordance with the inspiration of the individuals. Nothing came of them. To the last specimen, the newborn silkworms died of hunger. My renown as a quack must have suffered somewhat from this check. Was it really my fault? No, it was the fault of the silkworm, which remained faithful to its mulberry leaf.

It was therefore in nearly the certainty of non-fulfilment that I made my first attempts at rearing carnivorous larvae with a quarry which did not conform with the customary regimen. For conscience' sake, more or less perfunctorily, I endeavoured to achieve something that seemed to me bound to end in pitiful failure. Only the Bembex-wasps, which are plentiful in the sand of the neighbouring hills, might still afford me, without too prolonged a search, a few subjects on which to experiment. The Tarsal Bembex furnished me with what I wanted: larvae young enough to have still before them a long period of feeding and yet sufficiently developed to endure the trials of a removal.

These larva are exhumed with all the consideration which their delicate skin demands; a number of head of game are likewise unearthed intact, having been recently brought by the mother. They consist of various Diptera, including some Anthrax-flies. An old sardine-box, containing a layer of sifted sand and divided into compartments by paper partitions, receives my charges, who are isolated one from another. These Fly-eaters I propose to turn into Grasshopper-eaters; for their Bembex-diet I intend to substitute the diet of a Sphex or a Tachytes. To save myself tedious errands devoted to provisioning the refectory, I accept what good fortune offers me at the very threshold of my door. A green Locustid, with a short sabre bent into a reaping-hook, Phaneroptera falcata, is ravaging the corollae of my petunias. Now is the time to indemnify myself for the damage which she has caused me. I pick her young, half to three-quarters of an inch in length; and I deprive her of movement, without more ado, by crushing her head. In this condition she is served up to the Bembex-larvae in place of their Flies.

If the reader has shared my convictions of failure, convictions based on very logical motives, he will now share my profound surprise. The impossible becomes possible, the senseless becomes reasonable and the expected becomes the opposite of the real. The dish served on the Bembeces' table for the first time since Bembeces came into the world is accepted without any repugnance and consumed with every mark of satisfaction. I will here set down the detailed diary of one of my guests; that of the others would only be a repetition, save for a few variations.

2 AUGUST, 1883. – The larva of the Bembex, as I extract it from its burrow, is about half-developed. Around it I find only some scanty relics of its meals, consisting chiefly of Anthrax-wings, half-diaphanous and half-clouded. The mother would appear to have completed the victualling by fresh contributions, added day by day. I give the nurseling, which is an Anthrax-eater, a young Phaneroptera. The Locustid is attacked without hesitation. This profound change in the character of its victuals does not seem in the least to disturb the larva, which bites straight into the rich morsel with its mandibles and does not let go until it has exhausted it. Towards evening the drained carcase is replaced by another, quite fresh, of the same species but bulkier, measuring over three-quarters of an inch.

3 AUGUST. – Next day I find the Phaneroptera devoured. Nothing remains but the dry integuments, which are not dismembered. The entire contents have disappeared; the game has been emptied through a large opening made in the belly. A regular Grasshopper-eater could not have operated more skilfully. I replace the worthless carcase by two small Locustidae. At first the larva does not touch them, being amply sated with the copious meal of the day before. In the afternoon, however, one of the items is resolutely attacked.

4 AUGUST. – I renew the victuals, although those of the day before are not finished. For the rest, I do the same daily, so that my charge may constantly have fresh food at hand. High game might upset its stomach. My

Locustidae are not victims at the same time living and inert, operated upon according to the delicate method of the insects that paralyse their prey; they are corpses, procured by a brutal crushing of the head. With the temperature now prevailing, flesh soon becomes tainted; and this compels me frequently to renew the provisions in my sardine-box refectory. Two specimens are served up. One is attacked soon afterwards; and the larva clings to it assiduously.

5 AUGUST. – The ravenous appetite of the start is becoming assuaged. My supplies may well be too generous; and it might be prudent to try a little dieting after this Gargantuan good cheer. The mother certainly is more parsimonious. If all the family were to eat at the same rate as my guest, she would never be able to keep pace with their demands. Therefore, for reasons of health, this is a day of fasting and vigil.

6 AUGUST. – Supplies are renewed with two Phaneropterae. One is consumed entirely; the other is bitten into.

7 AUGUST. – today's ration is tasted and then abandoned. The larva seems uneasy. With its pointed mouth it explores the walls of its chamber. This sign denotes the approach of the time for making the cocoon.

8 AUGUST. – During the night the larva has spun its silken eel-trap. It is now encrusting it with grains of sand. Then follow, in due time, the normal phases of the metamorphosis. Fed on Locustidae, a diet unknown to its race, the larva passes through its several stages without any more difficulty than its brothers and sisters fed on Flies.

I obtained the same success in offering young Mantes for food. One of the larvae thus served would even incline me to believe that it preferred the new dish to the traditional diet of its race. Two Eristales, or Drone-flies, and a Praying Mantis an inch long composed its daily allowance. The Drone-flies are disdained from the first mouthful; and the Mantis, already tasted and apparently found excellent, causes the Fly to be completely forgotten. Is this an epicure's preference, due to the greater juiciness of the flesh? I am not in a position to say. At all events, the Bembex is not so infatuated with Fly as to refuse to abandon it for other game.

The failure which I foresaw has proved a magnificent success. It is fairly convincing, is it not? Without the evidence of experiment, what can we rely upon? Beneath the ruins of so many theories which appeared to be most solidly erected I should hesitate to admit that two and two make four if the facts were not before me. My argument had the most tempting probability on its side, but it had not the truth. As it is always possible to find reasons after the event in support of an opinion which one would not at first admit, I should now argue as follows:

The plant is the great factory in which are elaborated, with mineral materials, the organic principles which are the materials of life. Certain products are common to the whole vegetable series, but others, far less numerous, are prepared in special laboratories. Each genus, each species has its trademark. Here essential oils are manufactured; here alkaloids; here starches, fatty substances, resins, sugars, acids. Hence result special energies, which do not suit every herbivorous animal. It assuredly requires a stomach made expressly for the purpose to digest aconite, colchicum, hemlock or henbane; those who have not such a stomach could never endure a diet of that sort. Besides, the Mithridates fed on poison resist only a single toxin. The caterpillar of the Death's-head Hawk-moth, which delights in the solanin of the potato, would be killed by the acrid principle of the tithymals that form the food of the Spurge-caterpillar. The herbivorous larvae are therefore perforce exclusive in their tastes, because different genera of vegetables possess very different properties.

With this variety in the products of the plant, the animal, a consumer far more than a producer, contrasts the uniformity in its own products. The albumen in the egg of the Ostrich or the Chaffinch, the casein in the milk of the Cow or the Ass, the muscular flesh of the Wolf or the Sheep, the Screech-owl or the Field-mouse, the Frog or the earthworm: these remain albumen, casein or fibrin, edible if not eaten. Here are no excruciating condiments, no special acridities, no alkaloids fatal to any stomach other than that of the appointed consumer; so that animal food is not confined to one and the same eater. What does not man eat, from that delicacy of the arctic regions, soup made of Seal's blood and a scrap of Whale-blubber wrapped in a willow-leaf for a vegetable, to the Chinaman's fried silkworm or the Arab's dried Locust? What would he not eat, if he had not to overcome the repugnance dictated by habit rather than by actual necessity? The prey being uniform in its nutritive principles, the carnivorous larva ought to accommodate itself to any sort of game, above all if the new dish be not too great a departure from consecrated usage. Thus should I argue, with no less probability on my side, had I to begin all over again. But, as all our arguments have not the value of a single fact, I should be forced in the end to resort to experiment.

I did so the next year, on a larger scale and with a greater variety of subjects. I shrink from a continuous narrative of my experiments and of my personal education in this new art, where the failure of one day taught me the way to succeed on the morrow. It would be long and tedious. Enough if I briefly state my results and the conditions which must be fulfilled in order to run the delicate refectory as it should be run.

And, first, we must not dream of detaching the egg from its natural prey to lay it on another. The egg adheres pretty firmly, by its cephalic pole, to the quarry. To remove it from its place would inevitably jeopardize its future. I therefore let the larva hatch and acquire sufficient strength to bear the removal without peril. For that matter, my excavations most often provide me with my subjects in the form of larvae. I adopt for rearing purposes the larvae that are a quarter to a half developed. The others are too young and risky to handle, or too old and limited to a short period of artificial feeding.

Secondly, I avoid bulky heads of game, a single one of which would suffice for the whole growing-stage. I have already said and I here repeat how nice a matter it is to consume a victim which has to keep fresh for a couple of weeks and not to finish dying until it is almost entirely devoured. Death here leaves no corpse; when life is extinct, the body has disappeared, leaving only a shred of skin. Larvae with only one large prey have a special art of eating, a dangerous art, in which a clumsy bite would prove fatal. If bitten before the proper time at such a point, the victim becomes putrid, which promptly causes death by poisoning in the consumer. When diverted from its plan of attack, deprived of its clue, the larva is not always able to rediscover the lawful morsels in good time and is killed by the decomposition of its badly dissected prey. What will happen if the experimenter gives it a game to which it is not accustomed? Not knowing how to eat it according to rule, the larva will kill it; and by next day the victuals will have become so much toxic putrescence. I have already told how I found it impossible to rear the Two-banded Scolia on Oryctes-larvae, fastened down to deprive them of movement, or even on Ephippigers, paralysed by the Languedocian Sphex. In both cases the new diet was accepted without hesitation, a proof that it suited the nurseling; but in a day or two putrescence supervened and the Scolia perished on the fetid morsel. The method of preserving the Ephippiger, so well known to the Sphex, was unknown to my boarder; in this was enough to convert a delicious food into poison.

Even so did my other attempts miscarry wretchedly, attempts at feeding with the single dish consisting of one big head of game to replace the normal ration. Only one success is recorded in my notebooks, but that was so difficult that I would not undertake to obtain it a second time. I succeeded in feeding the larva of the Hairy Ammophila with an adult black Cricket, who was accepted as readily as the natural game, the caterpillar.

To avoid putrefaction of victuals which last overlong and are not consumed according to the method indispensable to their preservation, I employ small game, each piece of which can be finished by the larva at a single sitting, or at most in a single day. It matters little then that the victim is slashed and dismembered at random; decomposition has no time to seize upon its still quivering tissues. This is the procedure of those larvae which gulp down their food, snapping at random without distinguishing one part from another, such as the Bembex-larvae, which finish the Fly into which they have bitten before beginning another in the heap, or the Cerceris-larvae, which drain their Weevils methodically one after another. With the first strokes of the mandibles the victim broached may be mortally wounded. This is no disadvantage: a brief spell suffices to make use of the corpse, which is saved from putrefaction by being promptly consumed. Close beside it, the other victims, quite alive though motionless, await their respective turns and supply reserves of victuals which are always fresh.

I am too unskilful a butcher to imitate the Wasp and myself to resort to paralysis; moreover, the caustic liquid injected into the nerve-centres, ammonia in particular, would leave traces of smell or flavour which might put off my boarders. I am therefore compelled to deprive my insects of the power of movement by killing them outright. This makes it impracticable to provide a sufficiency of provisions beforehand in a single supply: while one item of the ration was being consumed the rest would spoil. One expedient alone remains to me, one which entails constant attendance: it is to renew the provisions each day. When all these conditions are fulfilled, the success of artificial feeding is still not without its difficulties; nevertheless, with a little care and above all plenty of patience, it is almost certain.

It was thus that I reared the Tarsal Bembex, which eats Anthrax-flies and other Diptera, on young Locustidae or Mantidae; the Silky Ammophila, whose diet consists chiefly of Measuring-worms, on small Spiders; the pot-making Pelopaeus, a Spider-eater, on tender Acridians; the Sand Cerceris, a passionate lover of Weevils, on Halicti; the Bee-eating Philanthus, which feeds exclusively on Hive-bees, on Eristales and other Flies. Without succeeding in my final aim, for reasons which I have just explained, I have seen the Two-banded Scolia feasting greedily on the grub of the Oryctes, which was substituted for that of the Cetonia, and putting up with an Ephippiger taken from the burrow of the Sphex; I have been present at the repast of three Hairy Ammophilae accepting with an excellent appetite the Cricket that replaced their caterpillar. One of them, as I have related, contrived to keep its ration fresh, which enabled it to reach its full development and to spin its cocoon.

These examples, the only ones to which my experiments have extended hitherto, seem to me sufficiently convincing to allow me to conclude that the carnivorous larva does not have exclusive tastes. The ration supplied to it by the mother, so monotonous, so limited in quality, might be replaced by others equally to its taste. Variety does not displease the larva; it does it as much good as uniformity; indeed, it would be of greater benefit to the race, as we shall see presently.

CHAPTER 8.

A DIG AT THE EVOLUTIONISTS.

***

To rear a caterpillar-eater on a skewerful of Spiders is a very innocent thing, unlikely to compromise the security of the State; it is also a very childish thing, as I hasten to confess, and worthy of the schoolboy who, in the mysteries of his desk, seeks as best he may some diversion from the fascinations of his exercise in composition. And I should not have undertaken these investigations, still less should I have spoken them, not without some satisfaction, if I had not discerned, in the results obtained in my refectory, a certain philosophic import, involving, so it seemed to me, the evolutionary theory.

It is assuredly a majestic enterprise, commensurate with man's immense ambitions, to seek to pour the universe into the mould of a formula and submit every reality to the standard of reason. The geometrician proceeds in this manner: he defines the cone, an ideal conception; then he intersects it by a plane. The conic section is submitted to algebra, an obstetrical appliance which brings forth the equation; and behold, entreated now in one direction, now in another, the womb of the formula gives birth to the ellipse, the hyperbola, the parabola, their foci, their radius vectors, their tangents, their normals, their conjugate axes, their asymptotes and the rest. It is magnificent, so much so that you are overcome by enthusiasm, even when you are twenty years old, an age hardly adapted to the austerities of mathematics. It is superb. You feel as if you were witnessing the creation of a world.

As a matter of fact, you are merely observing the same idea from different points of view, which are illumined by the successive phases of the transformed formula. All that algebra unfolds for our benefit was contained in the definition of the cone, but it was contained as a germ, under latent forms which the magic of the calculus converts into explicit forms. The gross value which our mind confided to the equation it returns to us, without loss or gain, in coins stamped with every sort of effigy. And here precisely is that which constitutes the inflexible rigour of the calculus, the luminous certainty before which every cultivated mind is forced to bow. Algebra is the oracle of the absolute truth, because it reveals nothing but what the mind had hidden in it under an amalgam of symbols. We put 2 and 2 into the machine; the rollers work and show us 4. That is all.

But to this calculus, all-powerful so long as it does not leave the domain of the ideal, let us submit a very modest reality: the fall of a grain of sand, the pendular movement of a hanging body. The machine no longer works, or does so only by suppressing almost everything that is real. It must have an ideal material point, an ideal rigid thread, an ideal point of suspension; and then the pendular movement is translated by a formula. But the problem defies all the artifices of analysis if the oscillating body is a real body, endowed with volume and friction; if the suspensory thread is a real thread, endowed with weight and flexibility; if the point of support is a real point, endowed with resistance and capable of deflection. So with other problems, however simple. The exact reality escapes the formula.

Yes, it would be a fine thing to put the world into an equation, to assume as the first principle a cell filled with albumen and by transformation after transformation to discover life under its thousand aspects as the geometrician discovers the ellipse and the other curves by examining his conic section. Yes, it would be magnificent and enough to add a cubit to our stature. Alas, how greatly must we abate our pretensions! The reality is beyond our reach when it is only a matter of following a grain of dust in its fall; and we would undertake to ascend the river of life and trace it to its source! The problem is a more arduous one than that which algebra declines to solve. There are formidable unknown quantities here, more difficult to decipher than the resistances, the deflections and the frictions of the pendulum. Let us eliminate them, that we may more easily propound the theory.

Very well; but then my confidence in this natural history which repudiates nature and gives ideal conceptions precedence over real facts is shaken. So, without seeking the opportunity, which is not my business, I take it when it presents itself; I examine the theory of evolution from every side; and, as that which I have been assured is the majestic dome of a monument capable of defying the ages appears to me to be no more than a bladder, I irreverently dig my pin into it.

Here is the latest dig. Adaptability to a varied diet is an element of wellbeing in the animal, a factor of prime importance for the extension and predominance of its race in the bitter struggle for life. The most unfortunate species would be that which depended for its existence on a diet so exclusive that no other could replace it. What would become of the Swallow if he required, in order to live, one particular Gnat, a single Gnat, always the same? When once this Gnat had disappeared - and the life of the Mosquito is not a long one - the bird would die of starvation. Fortunately for himself and for the happiness of our homes, the Swallow gulps them all down indiscriminately, together with a host of other insects that perform aerial ballets. What would become of the Lark were his gizzard able to digest only one seed, invariably the same? When the season for this seed was over - and the season is always a short one - the haunter of the furrows would perish.

Is not man's complaisant stomach, adapted to the largest variety of nourishment, one of his great zoological privileges? He is thus rendered independent of climates, seasons and latitudes. And the Dog: how is it that of all the domestic animals he alone is able to accompany us everywhere, even on the most arduous expeditions? The Dog again is omnivorous and therefore a cosmopolitan.

The discovery of a new dish, said Brillat-Savarin, is of greater importance to humanity than the discovery of a new planet. The aphorism is nearer to the truth than it appears to be in its humorous form. Certainly the man who was the first to think of crushing wheat, kneading flour and cooking the paste between two hot stones was more deserving than the discoverer of the two-hundredth asteroid. The invention of the potato is certainly as valuable as that of Neptune, glorious as the latter was. All that increases our alimentary resources is a discovery of the first merit. And what is true of man cannot be other than true of animals. The world belongs to the stomach which is independent of specialities. This truth is of the kind that has only to be stated to be proved.

Let us now return to our insects. If I am to believe the evolutionists, the various game-hunting Wasps are descended from a small number of types, which are themselves derived, by an incalculable number of concatenations, from a few amoebae, a few monera and lastly from the first clot of protoplasm which was casually condensed. Let us not go back as far as that; let us not plunge into the fogs where illusion and error too easily find a lurking-place. Let us consider a subject with exact limits to it; this is the only way to understand one another.

The Sphegidae are descended from a single type, which itself was already a highly-developed descendant and, like its successors, fed its family on prey. The close similarity in form, in colouring and, above all, in habits seem to refer the Tachytes to the same origin. This is ample; let us be satisfied with it. And now please tell me, what did this prototype of the Sphegidae hunt? Was its diet varied or uniform? If we cannot decide, let us examine the two cases.

The diet was varied. I heartily congratulate the first born of the Sphex-wasps. She enjoyed the most favourable conditions for leaving a prosperous offspring. Accommodating herself to any kind of prey not disproportionate to her strength, she avoided the dearth of a given species of game at this or that time and in this or that place; she always found the wherewithal to endow her family magnificently, they being, for that matter, fairly indifferent to the nature of the victuals, provided that these consisted of fresh insect-flesh, as the tastes of their cousins many times removed prove to this day. This matriarch of the Sphex clan bore within herself the best chances of assuring victory to her offspring in that pitiless fight for existence which eliminates the weakly and incapable and allows none but the strong and industrious to survive; she possessed an aptitude of great value which atavism could not fail to hand down and which her descendants, who are greatly interested in preserving this magnificent inheritance, must have permanently adopted and even accentuated from one generation to the next, from one branch, one offshoot, to another.

Instead of this unscrupulously omnivorous race, levying booty upon every kind of game, to its very great advantage, what do we see today? Each Sphex is stupidly limited to an unvarying diet; she hunts only one kind of prey, though her larva accepts them all. One will have nothing but the Ephippiger and must have a female at that; another will have nothing but the Cricket. This one hunts the Locust and nothing else; that one the Mantis and the Empusa. Yet another is addicted to the Grey Worm and another to the Looper.

Fools! How great was your mistake in allowing the wise eclecticism of your ancestress, whose relics now repose in the hard mud of some lacustrian stratum, to become obsolete! How much better would things be for you and yours! Abundance is assured; painful and often fruitless searches are avoided; the larder is crammed without being subject to the accidents of time, place and climate. When Ephippigers run short, you fall back upon Crickets; when there are no Crickets, you capture Grasshoppers. But no, my beautiful Sphex-wasps, you were not such fools as that. If in our days you are each confined to a standing family-dish, it is because your ancestress of the lacustrian schists never taught you variety.

Could she have taught you uniformity? Let us suppose that the Sphex of antiquity, a novice in the gastronomic art, prepared her potted meats with a single kind of game, no matter what. It was then her descendants who, subdivided into groups and constituted into so many distinct species by the slow travail of the centuries, realized that in addition to the ancestral fare there existed a host of other foods. Tradition being abandoned, there was nothing to guide their choice. They therefore tried a bit of everything in the way of insect game, at haphazard; and each time the larva, whose tastes alone had to be consulted, was satisfied with the food supplied, as it is today in the refectory provisioned by my care.

Every attempt led to the invention of a new dish, an important event, according to the masters, an inestimable resource for the family, who were thereby delivered from the menace of death and enabled to thrive over large areas whence the absence or rarity of a uniform game would have excluded it. And, after making use of a host of different viands in order to attain the culinary variety which is today adopted by the whole of the Sphex nation, lo and behold, each species confines itself to a single sort of game, outside which every specimen is obstinately refused, not at table, of course, but in the hunting-field! By your experiments, from age to age, to have discovered variety in diet; to have practised it, to the great advantage of your race, and to end up with uniformity, the cause of decadence; to have known the excellent and to repudiate it for the middling: oh, my Sphex-wasps, it would be stupid if the theory of evolution were correct!

To avoid insulting you and also from respect for common sense, I prefer therefore to believe that, if in our days you confine your hunting to a single kind of game, it is because you have never known any other. I prefer to believe that your common ancestress, your precursor, whether her tastes were simple or complex, is a pure chimera, for, if they were any relationship between you, having tested everything in order to arrive at the actual food of each species, having eaten everything and found it grateful to the stomach, you would now, from first to last, be unprejudiced consumers, omnivorous progressives. I prefer to believe, in short, that the theory of evolution is powerless to explain your diet. This is the conclusion drawn from the dining-room installed in my old sardine-box.

CHAPTER 9.

RATIONING ACCORDING TO SEX.

***

Considered in respect of quantity, the food sets us other problems, no less obscure. Those of us who devote ourselves assiduously to studying the customs of the game-hunting Wasps soon find our attention arrested by a very remarkable fact, at the time when our mind, refusing to be satisfied with sweeping generalities, which our indolence too readily makes shift with, seeks to enter as far as possible into the secret of the details, so curious and sometimes so important, as and when they become better-known to us. This fact, which has preoccupied me for many a long year, is the variable quantity of the provisions packed into the burrow as food for the larva.

Each species is scrupulously faithful to the diet of its ancestors. For more than a quarter of a century I have been exploring my district; and I have never known the diet to vary. today, as thirty years ago, each huntress must have the game which I first saw her pursuing. But, though the nature of the victuals is constant, the quantity is not so. In this respect the difference is so great that he would need to be a very superficial observer who should fail to perceive it on his first examination of the burrows. In the beginning, this difference, involving two, three, four times the quantity and more, perplexed me extremely and led me to the conclusions which I reject today.

Here, among the instances most familiar to me, are some examples of these variations in the number of victims provided for the larva, victims, of course, very nearly identical in size. In the larder of the Yellow-winged Sphex, after the victualling is completed and the house shut up, two or three Crickets are sometimes found and sometimes four. Stizus ruficornis, established in some vein of soft sandstone, places three Praying Mantes in one cell and five in another. Of the caskets fashioned by Amedeus' Eumenes out of clay and bits of stone, the more richly endowed contain ten small caterpillars, the more poorly furnished five. The Sand Cerceris will sometimes provide a ration of eight Weevils and sometimes one of twelve or even more. My notes abound in abstracts of this kind. It is unnecessary for the purpose in hand to quote them all. It will serve our object better if I give the detailed inventory of the Bee-eating Philanthus and of the Mantis-hunting Tachytes, considered especially with regard to the quantity of the victuals.

The slayer of Hive-bees is frequently in my neighbourhood; and I can obtain from her with the least trouble the greatest number of data. In September I see the bold filibuster flying from clump to clump of the pink heather pillaged by the Bee. The bandit suddenly arrives, hovers, makes her choice and swoops down. The trick is done: the poor worker, with her tongue lolling from her mouth in the death-struggle, is carried through the air to the underground den, which is often a very long way from the spot of the capture. The trickling of earthy refuse, on the bare banks, or on the slopes of footpaths, instantly reveals the dwellings of the ravisher; and, as the Philanthus always works in fairly populous colonies, I am able, by noting the position of the communities, to make sure of fruitful excavations during the forced inactivity of winter.

The sapping is a laborious task, for the galleries run to a great depth. Favier wields the pick and spade; I break the clods which he brings down and open the cells, whose contents - cocoons and remnants of provisions - I at once pour into a little screw of paper. Sometimes, when the larva is not developed, the stack of Bees is intact; more often the victuals have been consumed; but it is always possible to tell the number of items provided. The heads, abdomens and thoraxes, emptied of their fleshy substance and reduced to the tough outer skin, are easily counted. If the larva has chewed these overmuch, the wings at least are left; these are sapless organs which the Philanthus absolutely scorns. They are likewise spared by moisture, putrefaction and time, so much so that it is no more difficult to take an inventory of a cell several years old than one of a recent cell. The essential thing is not to overlook any of these tiny relics while placing them in the paper bag, amid the thousand incidents of the excavation. The rest of the work will be done in the study, with the aid of the lens, taking the remains heap by heap; the wings will be separated from the surrounding refuse and counted in sets of four. The result will give the amount of the provisions. I do not recommend this task to any one who is not endowed with a good stock of patience, nor above all to any one who does not start with the conviction that results of great interest are compatible with very modest means.

My inspection covers a total of one hundred and thirty-six cells, which are divided as in the table below:

2 cells each containing	1 Bee
52 cells each containing	2 Bees
36 cells each containing	3 Bees
36 cells each containing	4 Bees
9 cells each containing	5 Bees
1 cell containing	6 Bees

136	

The Mantis-hunting Tachytes consumes its heap of Mantes, the horny envelope included, without leaving any remains but scanty crumbs, quite insufficient to establish the number of items provided. After the meal is completed, any inventory of the rations becomes impossible. I therefore have recourse to the cells which still contain the egg or the very young larva and, above all, to those whose provisions have been invaded by a tiny parasitic Gnat, a Tachina, which drains the game without cutting it up and leaves the whole skin intact. Twenty-five larders, put to the count, give me the following result:

8 cells each containing	3 items
5 cells each containing	4 items
4 cells each containing	6 items
3 cells each containing	7 items
2 cells each containing	8 items
1 cell containing	9 items
1 cell containing	12 items
1 cell containing	16 items

25	

The predominant game is the Praying Mantis, green; next comes the Grey Mantis, ash-coloured. A few Empusae make up the total. The specimens vary in dimensions within fairly elastic limits: I measure some which are a third to a half inch long, averaging two-thirds to one inch long, and some which are two-fifths, averaging three quarters. I see pretty plainly that their number increases in proportion as their size diminishes, as though the Tachytes were seeking to make up for the smallness of the game by increasing the amount; none the less I find it quite impossible to detect the least equivalence by combining the two factors of number and size. If the huntress really estimates the provisions, she does so very roughly; her household accounts are not at all well kept; each head of game, large or small, must always count as one in her eyes.

Put on my guard, I look to see whether the honey-gathering Bees have a double service, like the game-hunting Wasps'. I estimate the amount of honeyed paste; I gauge the cups intended to contain it. In many cases the result resembles the first obtained: the abundance of provisions varies from one cell to another. Certain Osmiae (O. cornuta and O. tricornis feed their larvae on a heap of pollen-dust moistened in the middle with a very little disgorged honey. One of these heaps may be three or four times the size of some other in the same group of cells. If I detach from its pebble the nest of the Mason-bee, the Chalicodoma of the Walls, I see cells of large capacity, sumptuously provisioned; close beside these I see others, of less capacity, with victuals parsimoniously allotted. The fact is general; and it is right that we should ask ourselves the reason for these marked differences in the relative quantity of foodstuffs and for these unequal rations.

I at last began to suspect that this is first and foremost a question of sex. In many Bees and Wasps, indeed, the male and the female differ not only in certain details of internal or external structure - a point of view which does not affect the present problem - but also in length and bulk, which depend in a high degree on the quantity of food.

Let us consider in particular the Bee-eating Philanthus. Compared with the female, the male is a mere abortion. I find that he is only a third to half the size of the other sex, as far as I can judge by sight alone. To obtain exactly the respective quantities of substance, I should need delicate balances, capable of weighing down to a milligramme. My clumsy villager's scales, on which potatoes may be weighed to within a kilogramme or so, do not permit of this precision. I must therefore rely on the evidence of my sight alone, evidence, for that matter, which is amply sufficient in the present instance. Compared with his mate, the Mantis-hunting Tachytes is likewise a pigmy. We are quite astonished to see him pestering his giantess on the threshold of the burrows.

We observe differences no less pronounced of size - and consequently of volume, mass and weight - in the two sexes of many Osmiae. The differences are less emphatic, but are still on the same side, in the Cerceres, the Stizi, the Spheges, the Chalicodomae and many more. It is therefore the rule that the male is smaller than the female. There are of course some exceptions, though not many; and I am far from denying them. I will mention certain Anthidia where the male is the larger of the two. Nevertheless, in the great majority of cases the female has the advantage.

And this is as it should be. It is the mother, the mother alone, who laboriously digs underground galleries and chambers, kneads the plaster for coating the cells, builds the dwelling-house of cement and bits of grit, bores the wood and divides the burrow into storeys, cuts the disks of leaf which will be joined together to form honey-pots, works up the resin gathered in drops from the wounds in the pine-trees to build ceilings in the empty spiral of a Snail-shell, hunts the prey, paralyses it and drags it indoors, gathers the pollen-dust, prepares the honey in her crop, stores and mixes the paste. This severe labour, so imperious and so active, in which the insect's whole life is spent, manifestly demands a bodily strength which would be quite useless to the male, the amorous trifler. Thus, as a general rule, in the insects which carry on an industry the female is the stronger sex.

Does this pre-eminence imply more abundant provisions during the larval stage, when the insect is acquiring the physical growth which it will not exceed in its future development? Simple reflection supplies the answer: yes, the aggregate growth has its equivalent in the aggregate provisions. Though so slight a creature as the male Philanthus finds a ration of two Bees sufficient for his needs, the female, twice or thrice as bulky, will consume three to six at least. If the male Tachytes requires three Mantes, his consort's meal will demand a batch of something like ten. With her comparative corpulence, the female Osmia will need a heap of paste twice or thrice as great as that of her brother, the male.

All this is obvious; the animal cannot make much out of little.

Despite this evidence, I was anxious to enquire whether the reality corresponded with the previsions of the most elementary logic. Instances are not unknown in which the most sagacious deductions have been found to disagree with the facts. During the last few years, therefore, I have profited by my winter leisure to collect, from spots noted as favourable during the working-season, a few handfuls of cocoons of various Digger-wasps, notably of the Bee-eating Philanthus, who has just furnished us with an inventory of provisions. Surrounding these cocoons and thrust against the wall of the cell were the remnants of the victuals - wings, corselets, heads, wing-cases - a count of which enabled me to determine how many head of game had been provided for the larva, now enclosed in its silken abode. I thus obtained the correct list of provisions for each of the huntress' cocoons. On the other hand, I estimated the quantities of honey, or rather I gauged the receptacles, the cells, whose capacity is proportionate to the mass of the provisions stored. After making these preparations, registering the cells, cocoons and rations and putting all my figures in order, I had only to wait for the hatching-season to determine the sex.

Well, I found that logic and experiment were in perfect agreement. The Philanthus-cocoons with two Bees gave me males, always males; those with a larger ration gave me females. From the Tachytes-cocoons with double or treble that ration I obtained females. When fed upon four or five Nut-weevils, the Sand Cerceris was a male; when fed upon eight or ten, a female. In short, abundant provisions and spacious cells yield females; scanty provisions and narrow cells yield males. This is a law upon which I may henceforth rely.

At the stage which we have now reached a question arises, a question of major importance, touching the most nebulous aspect of embryogeny. How is it that the larva of the Philanthus, to take a particular case, receives three to five Bees from its mother when it is to become a female and not more than two when it is to become a male? Here the various head of game are identical in size, in flavour, in nutritive properties. The food-value is precisely in proportion to the number of items supplied, a helpful detail which eliminates the uncertainties wherein we might be left by the provision of game of different species and varying sizes. How is it, then, that a host of Bees and Wasps, of honey-gatherers as well as huntresses, store a larger or smaller quantity of victuals in their cells according as the nurselings are to become females or males?

The provisions are stored before the eggs are laid; and these provisions are measured by the needs of the sex of an egg still inside the mother's body. If the egg-laying were to precede the rationing, which occasionally takes place, as with the Odyneri, for example, we might imagine that the gravid mother enquires into the sex of the egg, recognizes it and stacks victuals accordingly. But, whether destined to become a male or a female, the egg is always the same; the differences - and I have no doubt that there are differences - are in the domain of the infinitely subtle, the mysterious, imperceptible even to the most practised embryogenist. What can a poor insect see - in the absolute darkness of its burrow, moreover - where science armed with optical instruments has not yet succeeded in seeing anything? And besides, even were it more discerning than we are in these genetic obscurities, its visual discernment would have nothing whereupon to practice. As I have said, the egg is laid only when the corresponding provisions are stored. The meal is prepared before the larva which is to eat it has come into the world. The supply is generously calculated by the needs of the coming creature; the dining-room is built large or small to contain a giant or a dwarf still germinating in the ovarian ducts. The mother, therefore, knows the sex of her egg beforehand.

A strange conclusion, which plays havoc with our current notions! The logic of the facts leads us to it directly. And yet it seems so absurd that, before accepting it, we seek to escape the predicament by another absurdity. We wonder whether the quantity of food may not decide the fate of the egg, originally sexless. Given more food and more room, the egg would become a female; given less food and less room, it would become a male. The mother, obeying her instincts, would store more food in this case and less in that; she would build now a large and now a small cell; and the future of the egg would be determined by the conditions of food and shelter.

Let us make every test, every experiment, down to the absurd: the crude absurdity of the moment has sometimes proved to be the truth of the morrow. Besides, the well-known story of the Hive-bee should make us wary of rejecting paradoxical suppositions. Is it not by increasing the size of the cell, by modifying the quality and quantity of the food, that the population of a hive transforms a worker larva into a female or royal larva? It is true that the sex remains the same, since the workers are only incompletely developed females.

The change is none the less miraculous, so much so that it is almost lawful to enquire whether the transformation may not go further, turning a male, that poor abortion, into a sturdy female by means of a plentiful diet. Let us therefore resort to experiment.

I have at hand some long bits of reed in the hollow of which an Osmia, the Three-horned Osmia, has stacked her cells, bounded by earthen partitions. I have related elsewhere how I obtain as many of these nests as I could wish for. When the reed is split lengthwise, the cells come into view, together with their provisions, the egg lying on the paste, or even the budding larva. Observations multiplied ad nauseam have taught me where to find the males and where the females in this apiary. The males occupy the forepart of the reed, the end next to the opening; the females are at the bottom, next to the knot which serves as a natural stopper to the channel. For the rest, the quantity of the provisions in itself points to the sex: for the females it is twice or thrice as great as for the males.

In the scantily-provided cells, I double or treble the ration with food taken from other cells; in the cells which are plentifully supplied, I reduce the portion to a half or a third. Controls are left: that is to say, some cells remain untouched, with their provisions as I found them, both in the part which is abundantly provided and in that which is more meagrely rationed. The two halves of the reed are then restored to their original position and firmly bound with a few turns of wire. We shall see, when the time comes, whether these changes increasing or decreasing the victuals have determined the sex.

Here is the result: the cells which at first were sparingly provided, but whose supplies were doubled or trebled by my artifice, contain males, as foretold by the original amount of victuals. The surplus which I added has not completely disappeared, far from it: the larva has had more than it needed for its evolution as a male; and, being unable to consume the whole of its copious provisions, it has spun its cocoon in the midst of the remaining pollen-dust. These males, so richly supplied, are of handsome but not exaggerated proportions; you can see that the additional food has profited them to some small extent.

The cells with abundant provisions, reduced to a half or a third by my intervention, contain cocoons as small as the male cocoons, pale, translucent and limp, whereas the normal cocoons are dark-brown, opaque and firm to the touch. These, we perceive at once, are the work of starved, anaemic weavers, who, failing to satisfy their appetite and having eaten the last grain of pollen, have, before dying, done their best with their poor little drop of silk. Those cocoons which correspond with the smallest allowance of food contain only a dead and shrivelled larva; others, in whose case the provisions were less markedly decreased, contain females in the adult form, but of very diminutive size, comparable with that of the males, or even smaller. As for the controls which I was careful to leave, they confirm the fact that I had males in the part near the orifice of the reed and females in the part near the knot closing the channel.

Is this enough to dispose of the very improbable supposition that the determination of the sex depends on the quantity of food? Strictly speaking, there is still one door open to doubt. It may be said that experiment, with its artifices, does not succeed in realizing the delicate natural conditions. To make short work of all objections, I cannot do better than have recourse to facts in which the experimenter's hand has not intervened. The parasites will supply us with these facts; they will show us how alien the quantity and even the quality of the food are from either specific or sexual characters. The subject of enquiry thus becomes double, instead of single as it was when I plundered one cell in my split reeds to enrich another. Let us follow this double current for a little while.

An Ammophila, the Silky Ammophila, which feeds on Looper caterpillars, has just been reared in my refectory on Spiders. Replete to the regulation point, it spins its cocoon. What will emerge from this? If the reader expects to see any modifications, caused by a diet which the species, left to itself, had never effected, let him be undeceived and that quickly. The Ammophila fed on Spiders is precisely the same as the Ammophila fed on caterpillars, just as man fed on rice is the same as man fed on wheat. In vain I pass my lens over the product of my art: I cannot distinguish it from the natural product; and I defy the most meticulous entomologist to perceive any difference between the two. It is the same with my other boarders who have had their diet altered.

I see the objection coming. The differences may be inappreciable, for my experiments touch only a first rung of the ladder. What would happen if the ladder were prolonged, if the offspring of the Ammophila fed on Spiders were given the same food generation after generation? These differences, at first imperceptible, might become accentuated until they grew into distinct specific characters; the habits and instincts might also change; and in the end the caterpillar-huntress might become a Spider-huntress, with a shape of her own. A species would be created, for, among the factors at work in the transformation of animals, the most important of all is incontestably the type of food, the nature of the thing wherewith the animal builds itself. All this is much more important than the trivialities which Darwin relies upon.

To create a species is magnificent in theory, so that we find ourselves regretting that the experimenter is not able to continue the attempt. But, once the Ammophila has flown out of the laboratory to slake her thirst at the flowers in the neighbourhood, just to try to find her again and induce her to entrust you with her eggs, which you would rear in the refectory, to increase the taste for Spiders from generation to generation! Merely to dream of it were madness. Shall we, in our helplessness, admit ourselves beaten by the evolutionary effects of diet? Not a bit of it! One experiment - and you could not wish for a more decisive - is continually in progress, apart from all artifices, on an enormous scale. It is brought to our notice by the parasites.

They must, we are told, have acquired the habit of living on others in order to save themselves work and to lead an easier life. The poor wretches have made a sorry blunder. Their life is of the hardest. If a few establish themselves comfortably, dearth and dire famine await most of the rest. There are some - look at certain of the Oil-beetles - exposed to so many chances of destruction that, to save one, they are obliged to procreate a thousand. They seldom enjoy a free meal.

Some stray into the houses of hosts whose victuals do not suit them; others find only a ration quite insufficient for their needs; others – and these are very numerous – find nothing at all. What misadventures, what disappointments do these needy creatures suffer, unaccustomed as they are to work! Let me relate some of their misfortunes, gleaned at random.

The Girdled Dioxys (D. cincta) loves the ample honey-stores of the Chalicodoma of the Pebbles. There she finds abundant food, so abundant that she cannot eat it all. I have already passed censure on this waste. Now a little Osmia (O. cyanoxantha, Perez) makes her nest in the Mason's deserted cells; and this Bee, a victim of her ill-omened dwelling, also harbours the Dioxys. This is a manifest error on the parasite's part. The nest of the Chalicodoma, the hemisphere of mortar on its pebble, is what she is looking for, to confide her eggs to it. But the nest is now occupied by a stranger, by the Osmia, a circumstance unknown to the Dioxys, who comes stealing up to lay her egg in the mother's absence. The dome is familiar to her. She could not know it better if she had built it herself. Here she was born; here is what her family wants. Moreover, there is nothing to arouse her suspicions: the outside of the home has not changed its appearance in any respect; the stopper of gravel and green putty, which later will form a violent contrast with its white front, is not yet constructed. She goes in and sees a heap of honey. To her thinking this can be nothing but the Chalicodoma's portion. We ourselves would be beguiled, in the Osmia's absence. She lays her eggs in this deceptive cell.

Her mistake, which is easy to understand, does not in any way detract from her great talents as a parasite, but it is a serious matter for the future larva. The Osmia, in fact, in view of her small dimensions, collects but a very scanty store of food: a little loaf of pollen and honey, hardly the size of an average pea. Such a ration is insufficient for the Dioxys. I have described her as a waster of food when her larva is established, according to custom, in the cell of the Mason-bee. This description no longer applies; not in the very least. Inadvertently straying to the Osmia's table, the larva has no excuse for turning up its nose; it does not leave part of the food to go bad; it eats up the lot without having had enough.

This famine-stricken refectory can give us nothing but an abortion. As a matter of fact, the Dioxys subjected to this niggardly test does not die, for the parasite must have a tough constitution to enable it to face the disastrous hazards which lie in wait for it; but it attains barely half its ordinary dimensions, which means one-eighth of its normal bulk. To see it thus diminished, we are surprised at its tenacious vitality, which enables it to reach the adult form in spite of the extreme deficiency of food. Meanwhile, this adult is still the Dioxys; there is no change of any kind in her shape or colouring. Moreover, the two sexes are represented; this family of pigmies has its males and females. Dearth and the farinaceous mess in the Osmia's cell has had no more influence over species or sex than abundance and flowing honey in the Chalicodoma's home.

The same may be said of the Spotted Sapyga, which, a parasite of the Three-pronged Osmia, a denizen of the bramble, and of the Golden Osmia, an occupant of empty Snail-shells, strays into the house of the Tiny Osmia, where, for lack of sufficient food, it does not attain half its normal size.

A Leucopsis inserts her eggs through the cement wall of our three Chalicodomae. I know her under two names. When she comes from the Chalicodoma of the Pebbles or Walls, whose opulent larva saturates her with food, she deserves by her large size the name of Leucopsis gigas, which Fabricius bestows upon her; when she comes from the Chalicodoma of the Sheds, she deserves no more than the name of L. grandis, which is all that Klug grants her. With a smaller ration "the giant" is to some degree diminished and becomes no more than "the large." When she comes from the Chalicodoma of the Shrubs, she is smaller still; and, if some nomenclator were to seek to describe her, she would no longer deserve to be called more than middling. From dimension 2 she has descended to dimension 1 without ceasing to be the same insect, despite the change of diet; and at the same time both sexes are present in the three nurselings, despite the variation in the quantity of victuals.

I obtain Anthrax sinuata from various bees' nests. When she issues from the cocoons of the Three-horned Osmia, especially the female cocoons, she attains the greatest development that I know of. When she issues from the cocoons of the Blue Osmia (O. cyanea, KIRB.), she is sometimes hardly one-third the length which the other Osmia gives her. And we still have the two sexes - that goes without saying - and still identically the same species.

Two Anthidia, working in resin, A. septemdentatum, LATR., and A. bellicosum, LEP., establish their domicile in old Snail-shells. The second harbours the Burnt Zonitis (Z. proeusta). Amply nourished this Meloe then acquires her normal size, the size in which she usually figures in the collections. A like prosperity awaits her when she usurps the provisions of Megachile sericans. But the imprudent creature sometimes allows itself to be carried away to the meagre table of the smallest of our Anthidia (A. scapulare, LATR.), who makes her nests in dry bramble-stems. The scanty fare makes a wretched dwarf of the offspring belonging to either sex, without depriving them of any of their racial features. We still see the Burnt Zonitis, with the distinctive sign of the species: the singed patch at the tip of the wing-cases.

And the other Meloidae - Cantharides, Cerocomae, Mylabres - to what inequalities of size are they not subject, irrespective of sex! There are some - and they are numerous - whose dimensions fall to a half, a third, a quarter of the regular dimensions. Among these dwarfs, these misbegotten ones, these victims of atrophy, there are females as well as males; and their smallness by no means cools their amorous ardour. These needy creatures, I repeat, have a hard life of it. Whence do they come, these diminutive Beetles, if not from dining-rooms insufficiently supplied for their needs? Their parasitical habits expose them to harsh vicissitudes. No matter: in dearth as well as in abundance the two sexes appear and the specific features remain unchanged.

It is unnecessary to linger longer over this subject. The demonstration is completed. The parasites tell us that changes in the quantity and quality of food do not lead to any transformation of species. Fed upon the larva of the Three-horned Osmia or of the Blue Osmia, Anthrax sinuata, whether of handsome proportions or a dwarf, is still Anthrax sinuata; fed upon the allowance of the Anthidium of the empty Snail-shells, the Anthidium of the brambles, the Megachile or doubtless many others, the Burnt Zonitis is still the Burnt Zonitis. Yet variation of diet ought to be a very potential factor in the problem of progress towards another form. Is not the world of living creatures ruled by the stomach? And the value of this factor is unity, changing nothing in the product.

The same parasites tell us - and this is the chief object of my digression - that excess or deficiency of nutriment does not determine the sex. So we are once more confronted with the strange proposition, which is now more positive than ever, that the insect which amasses provisions in proportion to the needs of the egg about to be laid knows beforehand what the sex of this egg will be. Perhaps the reality is even more paradoxical still. I shall return to the subject after discussing the Osmiae, who are very weighty witnesses in this grave affair.

CHAPTER 10.

THE BEE-EATING PHILANTHUS.

***

To meet among the Wasps, those eager lovers of flowers, a species that goes hunting more or less on its own account is certainly a notable event. That the larder of the grub should be provided with prey is natural enough; but that the provider, whose diet is honey, should herself make use of the captives is anything but easy to understand. We are quite astonished to see a nectar-drinker become a blood-drinker. But our astonishment ceases if we consider things more closely. The double method of feeding is more apparent than real: the crop which fills itself with sugary liquid does not gorge itself with game. The Odynerus, when digging into the body of her prey, does not touch the flesh, a fare absolutely scorned as contrary to her tastes; she satisfies herself with lapping up the defensive drop which the grub distils at the end of its intestine. This fluid no doubt represents to her some highly-flavoured beverage with which she seasons from time to time the staple diet fetched from the drinking-bar of the flowers, some appetizing condiment or perhaps - who knows? - some substitute for honey. Though the qualities of the delicacy escape me, I at least perceive that the Odynerus does not covet anything else. Once its jar is emptied, the larva is flung aside as worthless offal, a certain sign of a non-carnivorous appetite. Under these conditions, the persecutor of the Chrysomela ceases to surprise us by indulging in the crying abuse of a double diet.

We even begin to wonder whether other species may not be inclined to derive a direct advantage from the hunting imposed upon them for the maintenance of the family. The Odynerus' method of work, the splitting open of the anal stillroom, is too far removed from the obvious procedure to have many imitators; it is a secondary detail and impracticable with a different kind of game. But there is sure to be a certain variety in the direct means of utilizing the capture. Why, for instance, when the victim paralysed by the sting contains a delicious broth in some part of its stomach, should the huntress scruple to violate her dying prey and force it to disgorge without injuring the quality of the provisions? There must be those who rob the dead, attracted not by the flesh but by the exquisite contents of the crop.

In point of fact, there are; and they are even numerous. We may mention in the first rank the Wasp that hunts Hive-bees, the Bee-eating Philanthus (P. apivorus, LATR.). I long suspected her of perpetrating these acts of brigandage on her own behalf, having often surprised her gluttonously licking the Bee's honey-smeared mouth; I had an inkling that she did not always hunt solely for the benefit of her larvae. The suspicion deserved to be confirmed by experiment. Also, I was engaged in another investigation, which might easily be conducted simultaneously with the one suggested: I wanted to study, with all the leisure of work done at home, the operating methods employed by the different Hunting Wasps. I therefore made use, for the Philanthus, of the process of experimenting under glass which I roughly outlined when speaking of the Odynerus. It was even the Bee-huntress who gave me my first data in this direction. She responded to my wishes with such zeal that I believed myself to possess an unequalled means of observing again and again, even to excess, what is so difficult to achieve on the actual spot. Alas, the first-fruits of my acquaintance with the Philanthus promised me more than the future held in store for me! But we will not anticipate; and we will place the huntress and her game together under the bell-glass. I recommend this experiment to whoever would wish to see with what perfection in the art of attack and defence a Hunting Wasp wields the stiletto. There is no uncertainty here as to the result, there is no long wait: the moment when she catches sight of the prey in an attitude favourable to her designs, the bandit rushes forward and kills. I will describe how things happen.

I place under the bell-glass a Philanthus and two or three Hive-bees. The prisoners climb the glass wall, towards the light; they go up, come down again and try to get out; the vertical polished surface is to them a practicable floor. They soon quiet down; and the spoiler begins to notice her surroundings. The antennae are pointed forwards, enquiringly; the hind-legs are drawn up with a little quiver of greed in the tarsi; the head turns to right and left and follows the evolutions of the Bees against the glass. The miscreant's posture now becomes a striking piece of acting: you can read in it the fierce longings of the creature lying in ambush, the crafty waiting for the moment to commit the crime. The choice is made: the Philanthus pounces on her prey.

Turn by turn tumbling over and tumbled, the two insects roll upon the ground. The tumult soon abates; and the murderess prepares to strangle her capture. I see her adopt two methods. In the first, which is more usual than the other, the Bee is lying on her back; and the Philanthus, belly to belly with her, grips her with her six legs while snapping at her neck with her mandibles. The abdomen is now curved forward from behind, along the prostrate victim, feels with its tip, gropes about a little and ends by reaching the under part of the neck. The sting enters, lingers for a moment in the wound; and all is over. Without releasing her prey, which is still tightly clasped, the murderess restores her abdomen to its normal position and keeps it pressed against the Bee's.

In the second method, the Philanthus operates standing. Resting on her hind-legs and on the tips of her unfurled wings, she proudly occupies an erect attitude, with the Bee held facing her between her four front legs. To give the poor thing a position suited to receive the dagger-stroke, she turns her round and back again with the rough clumsiness of a child handling its doll. Her pose is magnificent to look at. Solidly planted on her sustaining tripod, the two hinder tarsi and the tips of the wings, she at last crooks her abdomen upwards and again stings the Bee under the chin. The originality of the Philanthus' posture at the moment of the murder surpasses the anything that I have hitherto seen.

The desire for knowledge in natural history has its cruel side. To learn precisely the point attacked by the sting and to make myself thoroughly acquainted with the horrible talent of the murderess, I have investigated more assassinations under glass than I would dare to confess. Without a single exception, I have always seen the Bee stung in the throat. In the preparations for the final blow, the tip of the abdomen may well come to rest on this or that point of the thorax or abdomen; but it does not stop at any of these, nor is the sting unsheathed, as can readily be ascertained. Indeed, once the contest is opened, the Philanthus becomes so entirely absorbed in her operation that I can remove the cover and follow every vicissitude of the tragedy with my pocket-lens.

After recognizing the invariable position of the wound, I bend back and open the articulation of the head. I see under the Bee's chin a white spot, measuring hardly a twenty-fifth of an inch square, where the horny integuments are lacking and the delicate skin is shown uncovered. It is here, always here, in this tiny defect in the armour, that the sting enters. Why is this spot stabbed rather than another? Can it be the only vulnerable point, which would necessarily determine the thrust of the lancet? Should any one entertain so petty a thought, I advise him to open the articulation of the corselet, behind the first pair of legs. He will there see what I see: the bare skin, quite as fine as under the neck, but covering a much larger surface. The horny breastplate offers no wider breach. If the Philanthus were guided in her operation solely by the question of vulnerability, it is here certainly that she ought to strike, instead of persistently seeking the narrow slit in the neck. The weapon would not need to hesitate and grope; it would obtain admission into the tissues offhand. No, the stroke of the lancet is not forced upon it mechanically: the assassin scorns the large defect in the corselet and prefers the place under the chin, for eminently logical reasons which we will now attempt to unravel.

Immediately after the operation I take the Bee from the Philanthus. What strikes me is the sudden inertia of the antennae and the mouthparts, organs which in the victims of most of the Hunting Wasps continue to move for so long a time. There are here not any of the signs of life to which I have been accustomed in my old studies of insect paralysis: the antennary threads waving slowly to and fro, the palpi quivering, the mandibles opening and closing for days, weeks and months on end. At most, the tarsi tremble for a minute or two; that constitutes the whole death-struggle. Complete immobility ensues. The inference drawn from this sudden inertia is inevitable: the Wasp has stabbed the cervical ganglia. Hence the immediate cessation of movement in all the organs of the head; hence the real instead of the apparent death of the Bee. The Philanthus is a butcher and not a paralyser.

This is one step gained. The murderess chooses the under part of the chin as the point attacked in order to strike the principal nerve-centres, the cephalic ganglia, and thus to do away with life at one blow. When this vital seat is poisoned by the toxin, death is instantaneous. Had the Philanthus' object been simply to effect paralysis, the suppression of locomotor movements, she would have driven her weapon into the flaw in the corselet, as the Cerceres do with the Weevils, who are much more powerfully armoured than the Bee. But her intention is to kill outright, as we shall see presently; she wants a corpse, not a paralytic patient. This being so, we must agree that her operating-method is supremely well-inspired: our human murderers could achieve nothing more thorough or immediate.

We must also agree that her attitude when attacking, an attitude very different from that of the paralysers, is infallible in its death-dealing efficacy. Whether she deliver her thrust lying on the ground or standing erect, she holds the Bee in front of her, breast to breast, head to head. In this posture all that she need do is to curve her abdomen in order to reach the gap in the neck and plunge the sting with an upward slant into her captive's head. Suppose the two insects to be gripping each other in the reverse attitude, imagine the dirk to slant slightly in the opposite direction; the results would be absolutely different and the sting, driven downwards, would pierce the first thoracic ganglion and produce merely partial paralysis. What skill, to sacrifice a wretched Bee! In what fencing-school was the slayer taught her terrible upward blow under the chin?

If she learnt it, how is it that her victim, such a past mistress in architecture, such an adept in socialistic polity, has so far learnt no corresponding trick to serve in her own defence? She is as powerful as her executioner; like the other, she carries a rapier, an even more formidable one and more painful, at least to my fingers. For centuries and centuries the Philanthus has been storing her away in her cellars; and the poor innocent meekly submits, without being taught by the annual extermination of her race how to deliver herself from the aggressor by a well-aimed thrust. I despair of ever understanding how the assailant has acquired her talent for inflicting sudden death, when the assailed, who is better-armed and quite as strong, wields her dagger anyhow and therefore ineffectively. If the one has learnt by prolonged practice in attack, the other should also have learnt by prolonged practice in defence, for attack and defence possess a like merit in the fight for life. Among the theorists of the day, is there one clear-sighted enough to solve the riddle for us?

If so, I will take the opportunity of putting to him a second problem that puzzles me: the carelessness, nay, more, the stupidity of the Bee in the presence of the

Philanthus. You would be inclined to think that the victim of persecution, learning gradually from the misfortunes suffered by her family, would show distress at the ravisher's approach and at least attempt to escape. In my cages I see nothing of the sort. Once the first excitement due to incarceration under the bell-glass or the wire-gauze cover has passed, the Bee seems hardly to trouble about her formidable neighbour. I see one side by side with the Philanthus on the same honeyed thistle-head: assassin and future victim are drinking from the same flask. I see some one who comes heedlessly to enquire who that stranger can be, crouching in wait on the table. When the spoiler makes her rush, it is usually at a Bee who meets her halfway, and, so to speak, flings herself into her clutches, either thoughtlessly or out of curiosity. There is no wild terror, no sign of anxiety, no tendency to make off. How comes it that the experience of the ages, that experience which, we are told, teaches the animal so many things, has not taught the Bee the first element of apiarian wisdom: a deep-seated horror of the Philanthus? Can the poor wretch take comfort by relying on her trusty dagger? But she yields to none in her ignorance of fencing; she stabs without method, at random. However, let us watch her at the supreme moment of the killing.

When the ravisher makes play with her sting, the Bee does the same with hers and furiously. I see the needle now moving this way or that way in space, now slipping, violently curved, along the murderess' convex surface. These sword-thrusts have no serious results. The manner in which the two combatants are at grips has this effect, that the Philanthus' abdomen is inside and the Bee's outside. The latter's sting therefore finds under its point only the dorsal surface of the foe, a convex, slippery surface and so well armoured as to be almost invulnerable. There is here no breach into which the weapon can slip by accident; and so the operation is conducted with absolute surgical safety, notwithstanding the indignant protests of the patient.

After the fatal stroke has been administered, the murderess remains for a long time belly to belly with the dead, for reasons which we shall shortly perceive. There may now be some danger for the Philanthus. The attitude of attack and defence is abandoned; and the ventral surface, more vulnerable than the other, is within reach of the sting. Now the deceased still retains the reflex use of her weapon for a few minutes, as I learnt to my cost. Having taken the Bee too early from the bandit and handling her without suspecting any risk, I received a most downright sting. Then how does the Philanthus, in her long contact with the butchered Bee, manage to protect herself against that lancet, which is bent upon avenging the murder? Is there any chance of a commutation of the death-penalty? Can an accident ever happen in the Bee's favour? Perhaps.

One incident strengthens my faith in this perhaps. I had placed four Bees and as many Eristales under the bell-glass at the same time, with the object of estimating the Philanthus' entomological knowledge in the matter of the distinction of species. Reciprocal quarrels break out in the mixed colony. Suddenly, in the midst of the fray, the killer is killed. She tumbles over on her back, she waves her legs; she is dead. Who struck the blow? It was certainly not the excitable but pacific Drone-fly; it was one of the Bees, who struck home by accident during the thick of the fight. Where and how? I cannot tell. The incident occurs only once in my notes, but it throws a light upon the question. The Bee is capable of withstanding her adversary; she can then and there slay her would-be slayer with a thrust of the sting. That she does not defend herself to better purpose, when she falls into her enemy's clutches, is due to her ignorance of fencing and not to the weakness of her weapon. And here again arises, more insistently than before, the question which I asked above: how is it that the Philanthus has learnt for offensive what the Bee has not learnt for defensive purposes? I see but one answer to the difficulty: the one knows without having learnt; the other does not know because she is incapable of learning.

Let us now consider the motives that induce the Philanthus to kill her Bee instead of paralysing her. When the crime has been perpetrated, she manipulates her dead victim without letting go of it for a moment, holding its belly pressed against her own six legs. I see her recklessly, very recklessly, rooting with her mandibles in the articulation of the neck, sometimes also in the larger articulation of the corselet, behind the first pair of legs, an articulation of whose delicate membrane she is perfectly well aware, even though, when using her sting, she did not take advantage of this point, which is the most readily accessible of all. I see her rough-handling the Bee's belly, squeezing it against her own abdomen, crushing it in the press. The recklessness of the treatment is striking; it shows that there is no need for keeping up precautions. The Bee is a corpse; and a little hustling here and there will not deteriorate its quality, provided there be no effusion of blood. In point of fact, however rough the handling, I fail to discover the slightest wound.

These various manipulations, especially the squeezing of the neck, at once bring about the desired results: the honey in the crop mounts to the Bee's throat. I see the tiny drops spurt out, lapped up by the glutton as soon as they appear. The bandit greedily, over and over again, takes the dead insect's lolling, sugared tongue into her mouth; then she once more digs into the neck and thorax, subjecting the honey-bag to the renewed pressure of her abdomen. The syrup comes and is instantly lapped up and lapped up again. In this way the contents of the crop are exhausted in small mouthfuls, yielded one at a time. This odious meal at the expense of a corpse's stomach is taken in a sybaritic attitude; the Philanthus lies on her side with the Bee between her legs. The atrocious banquet sometimes lasts for half an hour or longer. At last the drained Bee is discarded, not without regret, it seems, for from time to time I see the manipulation renewed. After taking a turn round the top of the bell-jar, the robber of the dead returns to her prey and squeezes it, licking its mouth until the last trace of honey has disappeared.

This frenzied passion of the Philanthus for the Bee's syrup is declared in yet another fashion. When the first victim has been sucked dry, I slip under the glass a second victim, which is promptly stabbed under the chin and then subjected to pressure to extract the honey. A third follows and undergoes the same fate without satisfying the bandit. I offer a fourth and a fifth. They are all accepted. My notes mention one Philanthus who in front of my eyes sacrificed six Bees in succession and squeezed out their crops in the regulation manner. The slaughter came to an end not because the glutton was sated but because my functions as a purveyor were becoming rather difficult: the dry month of August causes the insects to avoid my harmas, which at this season is denuded of flowers. Six crops emptied of their honey: what an orgy! And even then the ravenous creature would very likely not have scorned a copious additional course, had I possessed the means of supplying it!

There is no reason to regret this break in the service; the little that I have said is more than enough to prove the singular characteristics of the Bee-slayer. I am far from denying that the Philanthus has an honest means of earning her livelihood; I find her working on the flowers as assiduously as the other Wasps, peacefully drawing her honeyed beakers. The males even, possessing no lancet, know no other manner of refreshment. The mothers, without neglecting the table d'hote of the flowers, support themselves by brigandage as well. We are told of the Skua, that pirate of the seas, that he swoops down upon the fishing birds, at the moment when they rise from the water with a capture. With a blow of the beak delivered in the pit of the stomach he makes them give up their prey, which is caught by the robber in midair. The despoiled bird at least gets off with nothing worse than a contusion at the base of the throat. The Philanthus, a less scrupulous pirate, pounces on the Bee, stabs her to death and makes her disgorge in order to feed upon her honey.

I say feed and I do not withdraw the word. To support my statement I have better reasons than those set forth above. In the cages in which various Hunting Wasps, whose stratagems of war I am engaged in studying, are waiting till I have procured the desired prey - not always an easy thing - I have planted a few flower-spikes, a thistle-head or two, on which are placed drops of honey renewed at need. Here my captives come to take their meals. With the Philanthus, the provision of honeyed flowers, though favourably received, is not indispensable. I have only to let a few live Bees into her cage from time to time. Half a dozen a day is about the proper allowance. With no other food than the syrup extracted from the slain, I keep my insects going for a fortnight or three weeks.

It is as plain as a pikestaff: outside my cages, when the opportunity offers, the Philanthus must also kill the Bee on her own account. The Odynerus asks nothing from the Chrysomela but a mere condiment, the aromatic juice of the rump; the other extracts from her victim an ample supplement to her victuals, the crop full of honey. What a hecatomb of Bees must not a colony of these freebooters make for their personal consumption, not to mention the stored provisions! I recommend the Philanthus to the signal vengeance of our Bee-masters.

Let us go no deeper into the first causes of the crime. Let us accept things as we know them for the moment, with their apparent or real atrocity. To feed herself, the Philanthus levies tribute on the Bee's crop. Having made sure of this, let us consider the bandit's method more closely. She does not paralyse her capture according to the rites customary among the Hunting Wasps; she kills it. Why kill it? If the eyes of our understanding be not closed, the need for sudden death is clear as daylight. The Philanthus proposes to obtain the honeyed broth without ripping up the Bee, a proceeding which would damage the game when it is hunted on behalf of the larvae, without resorting to the murderous extirpation of the crop. She must, by able handling, by skilful pressure, make the Bee disgorge, she must milk her, in a manner of speaking. Suppose the Bee stung behind the corselet and paralysed. That deprives her of her power of locomotion, but not of her vitality. The digestive organs in particular retain or very nearly retain their normal energy, as is proved by the frequent excretions that take place in the paralysed prey, so long as the intestine is not empty, as is proved above all by the victims of the Languedocian Sphex, those helpless creatures which I used to keep alive for forty days on end with a soup consisting of sugar and water. It is absurd to hope, without therapeutic means, without a special emetic, to coax a sound stomach into emptying its contents. The stomach of the Bee, who is jealous of her treasure, would lend itself to the process even less readily than another. When paralysed, the insect is inert; but there are always internal energies and organic forces which will not yield to the manipulator's pressure. The Philanthus will nibble at the throat and squeeze the sides in vain: the honey will not rise to the mouth so long as a vestige of life keeps the crop closed.

Things are different with a corpse. The tension is relaxed, the muscles become slack, the resistance of the stomach ceases and the bag of honey is emptied by the robber's vigorous pressure. You see, therefore, that the Philanthus is expressly obliged to inflict a sudden death, which will do away at once with the elasticity of the organs. Where is the lightning stroke to be delivered? The slayer knows better than we do, when she sticks the Bee under the chin. The cerebral ganglia are reached through the little hole in the neck and death ensues immediately.

The relation of these acts of brigandage cannot satisfy my distressing habit of following each reply obtained with a fresh question, until the granite wall of the unknowable rises before me. If the Philanthus is an expert in killing Bees and emptying crops swollen with honey, this cannot be merely an alimentary resource, especially when, in common with the others, she has the banqueting-hall of the flowers. I cannot accept her atrocious talent as inspired merely by the craving for a feast obtained at the expense of an empty stomach. Something certainly escapes us: the why and wherefore of that crop drained dry. A creditable motive may lie hidden behind the horrors which I have related. What is it?

Any one can understand the vagueness of the observer's mind when he first asks himself this question. The reader is entitled to be treated with consideration. I will spare him the recital of my suspicions, my gropings and my failures and will come straight to the results of my long investigation. Everything has its harmonious reason for existence. I am too fully persuaded of this to believe that the Philanthus pursues her habit of profaning corpses solely to satisfy her greed. What does the emptied crop portend? May it not be that . . .? Why, yes . . . After all, who knows? . . . Let us try along these lines.

The mother's first care is the welfare of the family. So far, we have seen the Philanthus hunting only for her stomach's sake; let us watch her hunting as a mother. Nothing is easier than to distinguish the two performances. When the Wasp wants a few good mouthfuls and nothing more, she scornfully abandons the Bee after picking her crop. The Bee is to her a worthless remnant, which will shrivel where it lies and be dissected by the Ants. If, on the other hand, she wants to stow away the Bee as a provision for her larvae, she clasps her in her two intermediate legs and, walking on the other four, goes round and round the edge of the bell-glass, seeking for an outlet through which to fly off with her prey. When she recognizes the circular track as impossible, she climbs up the sides, this time holding the Bee by the antennae with her mandibles and clinging to the polished and perpendicular surface with her six feet. She reaches the top of the glass, stays for a little while in the hollow of the knob at the top, returns to the ground, resumes her circling and her climbing and does not decide to relinquish her Bee until she has stubbornly attempted every means of escape. This persistence on her part to retain her hold on the cumbrous burden tells us pretty plainly that the game would go straight to the cells if the Philanthus had her liberty.

Well, these Bees intended for the larvae are stung under the chin like the others; they are real corpses; they are manipulated, squeezed, drained of their honey exactly as the others are. In all these respects, there is no difference between the hunt conducted to provide food for the larvae and the hunt conducted merely to gratify the mother's appetite.

As the worries of captivity might well be the cause of a few anomalies in the insect's actions, I felt that I ought to enquire how things happen in the open. I lay in wait near some colonies of Philanthi, for longer perhaps than the question deserved, as it had already been settled by what had happened under glass. My tedious watches were rewarded from time to time. Most of the huntresses returned home immediately, with the Bee under their abdomen; some halted on the brambles hard by; and here I saw them squeezing the dead Bee and making her disgorge the honey, which was greedily lapped up. After these preliminaries the corpse was stored. Every doubt is therefore removed: the provisions of the larva are first carefully drained of their honey.

Since we are on the spot, let us prolong our stay and enquire into the customs of the Philanthus in a state of liberty. Serving dead prey, which goes bad in a few days, the Bee-huntress cannot adopt the method of certain insects which paralyse a number of separate heads of game and fill the cell with provisions, completing the ration before laying the egg. She needs the method of the Bembex, whose larva receives the necessary nourishment at intervals, as it grows larger. The facts confirm this deduction. Just now I described as tedious my watches near the colonies of the Philanthi. They were tedious in fact, even more so perhaps than those which the Bembeces used to inflict upon me in the old days. Outside the burrows of the Great Cerceris and other Weevil-lovers, outside those of the Yellow-winged Sphex, the Cricket-slayer, there is plenty of distraction, thanks to the bustling movement of the hamlet. The mother has hardly come back home before she goes out again, soon returning laden with a new prey and once more setting out upon the chase. The going and coming is repeated at close intervals until the warehouse is full.

The burrow of the Philanthus is far from showing any such animation, even in a populous colony. In vain were my watches prolonged for whole mornings or afternoons; it was but very rarely that the mother whom I had seen go in with a Bee came out again for a second expedition. Two captures at most by the same huntress was all that I was able to see during my long vigils. Feeding from day to day involves this deliberation. Once the family is supplied with a sufficient ration for the moment, the mother suspends her hunting-trips until further need arises and occupies herself with mining-work in her underground house. Cells are dug; I see the rubbish gradually pushed up to the surface. Beyond this there is not a sign of activity; it is as though the burrow were deserted.

The inspection of the site is no easy matter. The shaft descends to a depth of nearly three feet in a compact soil, either vertically or horizontally. The spade and pick, wielded by stronger but less expert hands than mine, are indispensable, for which reason the process of excavation is far from satisfying me fully. At the end of this long tunnel, which the straw which I use for sounding despairs of ever reaching, the cells are at last encountered, oval cavities with a horizontal major axis. Their number and general arrangement escape me.

Some of them already contain the cocoon, which is slender and semitransparent, like those of the Cerceris, and, like them, suggests the shape of certain homoeopathic phials, with oval bellies surmounted by a tapering neck. The cocoon is fastened to the end of the cell by the tip of this neck, which is darkened and hardened by the larva's excrement; it has no other support. It looks like a short club fixed by the end of the handle along the horizontal axis of the nest. Other cells contain the larva in a more or less advanced stage. The grub is munching the last morsel served to it, with the scraps of the victuals already consumed lying around it. Others lastly show me a Bee, one only, still untouched and bearing an egg laid on her breast. This is the first partial ration; the others will come as and when the grub grows larger. My anticipations are thus confirmed:

following the example of the Bembeces, the Fly-killers, the Philanthus, the Bee-killer, lays her egg on the first piece warehoused and at intervals adds to her nurselings' repast.

The problem of the dead game is solved. There remains this other problem, one of incomparable interest: why are the Bees robbed of their honey before being served to the larvae? I have said and I say again that the killing and squeezing cannot be explained and excused simply by reference to the Philanthus' love of gormandizing. Robbing the worker of her booty is nothing out of the way: we see it daily; but cutting her throat in order to empty her stomach is going beyond a joke. And, as the Bees packed away in the cellar are squeezed dry just as much as the others, the thought occurs to my mind that a rumpsteak with jam is not to everybody's liking and that the game stuffed with honey might well be a distasteful or even unwholesome dish for the Philanthus' larvae. What will the grub do when, sated with blood and meat, it finds the Bee's honey-bag under its mandibles and especially when, nibbling at random, it rips open the crop and spoils its venison with syrup? Will it thrive on the mixture? Will the little ogre pass without repugnance from the gamy flavour of a carcase to the scent of flowers? A blunt statement or denial would serve no purpose. We must see. Let us see.

I rear some young Philanthus-grubs, already waxing large; but, instead of supplying them with the prey taken from the burrows, I give them game of my own catching, game replete with nectar from the rosemaries. My Bees, whom I kill by crushing their heads, are readily accepted; and I at first see nothing that corresponds with my suspicions. Then my nurselings languish, disdain their food, give a careless bite here and there and end by perishing, from the first to the last, beside their unfinished victuals. All my attempts miscarry: I do not once succeed in rearing my larvae to the stage of spinning the cocoon. And yet I am no novice in the functions of a foster-father. How many pupils have not passed through my hands and reached maturity in my old sardine-boxes as comfortably as in their natural burrows!

I will not draw rash conclusions from this check; I am conscientious enough to ascribe it to another cause. It may be that the atmosphere of my study and the dryness of the sand serving as a bed have had a bad effect on my charges, whose tender skins are accustomed to the warm moisture of the subsoil. Let us therefore try another expedient.

It is hardly feasible to decide positively by the methods which I have been following whether the honey is or is not repugnant to the grubs of the Philanthus. The first mouthfuls consist of meat; and then nothing particular occurs: it is the natural diet. The honey is met with later, when the morsel has been largely bitten into. If hesitation and lack of appetite are displayed at this stage, they come too late in the day to be conclusive: the larva's discomfort may be due to other, known or unknown, causes. The thing to do would be to offer the grub honey from the first, before artificial rearing has affected its appetite. It is useless, of course, to make the attempt with pure honey: no carnivorous creature would touch it, though it were starving. The jam-sandwich is the only device favourable to my plans, a meagre jam-sandwich, that is to say, the dead Bee lightly smeared or varnished with honey by means of a camel's-hair pencil.

Under these conditions, the problem is solved with the first few mouthfuls. The grub that has bitten into the honeyed prey draws back in disgust, hesitates a long time and then, urged by hunger, begins again, tries this side and that and ends by refusing to touch the dish. For a few days it pines away on top of its almost intact provisions; then it dies. All that are subjected to this regimen succumb. Do they merely perish of inanition in the presence of an unaccustomed food, which revolts their appetite, or are they poisoned by the small quantity of honey absorbed with the early mouthfuls? I cannot tell. The fact remains that, whether poisonous or repugnant, the Bee in the state of bread and jam is death to them; and this result explains, more clearly than the unfavourable circumstance of my former experiment, my failures with the Bee that had not been made to disgorge.

This refusal to touch the unwholesome or distasteful honey is connected with principles of nutrition which are too general to constitute a gastronomic peculiarity of the Philanthus. The other carnivorous larvae, at least in the order of the Hymenoptera, are bound to share it. Let us try. We will go to work as before. I unearth the larvae when they have attained a medium size, to avoid the weakness of infancy; I take away the natural provisions, smear the carcases separately with honey and, when this is done, restore its victuals to each of the grubs. I had to make a choice: not every subject was equally suited to my experiments. I must reject the larvae which are fed on one fat joint, such as those of the Scolia. The grub in fact attacks its prey at a determined point, dips its head and neck into the insect's body, rooting skilfully in the entrails to keep the game fresh until the end of the meal, and does not withdraw from the breach until the whole skin is emptied of its contents.

To make it let go with the object of coating the inside of the venison with honey had two drawbacks: I should be compromising the lingering vitality which saves the insect that is being devoured from going bad and, at the same time, I should be disturbing the delicate art of the devouring insect, which, if removed from the lode which it was working, would no longer be able to recover it or to distinguish between the lawful and the unlawful morsels. The larva of the Scolia, consuming its Cetonia-grub, has taught us all that we want to know on this subject in my earlier volume. The only acceptable larvae are those supplied with a heap of small insects, which are attacked without any special art, dismembered at random and eaten up quickly. Among these I have tested such as chance threw in my way: those of various Bembeces, all fed on Flies, those of the Palarus, whose bill of fare consists of a very large assortment of Hymenoptera; those of the Tarsal Tachytes, supplied with young Locusts; those of the Nest-building Odynerus, furnished with Chrysomela-grubs; those of the Sand Cerceris, endowed with a pinch of Weevils. A goodly variety, as you see, of consumers and consumed. Well, to all of these the seasoning with honey proved fatal. Whether poisoned or disgusted, they all died in a few days.

A strange result indeed! Honey, the nectar of the flowers, the sole diet of the Bee-tribe in both its forms and the sole resource of the Wasp in her a adult form, is to the larvae of the latter an object of insurmountable repugnance and probably a toxic dish. Even the transformation of the nymphosis surprises me less than this inversion of the appetite. What happens in the insect's stomach to make the adult seek passionately what the youngster refused lest it should die? This is not a question of organic debility unable to endure a too substantial, too hard, too highly spiced dish. The grub that gnaws the Cetonia-larva, that generous piece of butcher's meat; the glutton that crunches its batch of tough Locusts; the one that battens on nitrobenzine-flavoured game: they certainly own unfastidious gullets and accommodating stomachs. And these robust eaters allow themselves to die of hunger or digestive troubles because of a drop of syrup, the lightest food imaginable, suited to the weakness of extreme youth and a feast for the adult besides! What a gulf of obscurity in the stomach of a wretched grub!

These gastronomical researches called for a counter-experiment. The carnivorous larva is killed by honey. Conversely, is the mellivorous larva killed by animal food? Reservations are needful here, as in the previous tests. We should be courting a flat refusal if we offered a pinch of Locusts to the larvae of the Anthophora or the Osmia, for instance. The honey-fed insect would not bite into it. There would be no use whatever in trying. We must find the equivalent of the jam-sandwich aforesaid; in other words, we must give the larva its natural fare with a mixture of animal food. The addition made by my artifices shall be albumen, as found in the egg of the Hen, albumen the isomer of fibrin, which is the essential factor in any form of prey.

On the other hand, the Three-horned Osmia lends herself most admirably to my plans, because of her dry honey, consisting for the greater part of floury pollen. I therefore knead this honey with albumen, graduating the dose until its weight largely exceeds that of the flour. In this way I obtain pastes of different degrees of consistency, but all firm enough to bear the larva without danger of immersion. With too fluid a mixture there would be a risk of death by drowning. Lastly I install a moderately-developed larva on each of my albuminous cakes.

The dish of my inventing does not incite dislike: far from it. The grubs attack it without hesitation and consume it with every appearance of the usual appetite. Things could not go better if the food had not been altered by my culinary recipes. Everything goes down, including the morsels in which I feared that I had overdone the addition of albumen. And – an even more important point – the Osmia-larvae fed in this manner attain their normal dimensions and spin their cocoons, from which adult insects issue in the following year. Notwithstanding the albuminous regimen, the cycle of the evolution is achieved without impediment.

What are we to conclude from all this? I feel greatly embarrassed. Omne vivum ex ovo, the physiologists tell us. Every animal is carnivorous, in its first beginnings: it is formed and nourished at the cost of its egg, in which albumen predominates. The highest, the mammal, adheres to this diet for a long time: it has its mother's milk, rich in casein, another isomer of albumen. The gramnivorous nestling is first fed on grubs, which are better adapted to the niceties of its stomach; many of the minutest newborn creatures, being at once left to their own devices, take to animal food. In this way the original method of nourishment is continued for all alike: the method which allows flesh to be made from flesh and blood from blood, with no chemical process beyond the simplest modification. At maturity, when the stomach has acquired its full strength, vegetable food is adopted, involving a more complicated chemistry but easier to obtain. Milk is followed by fodder, worms by seeds, the prey in the burrow by the nectar of the flowers.

This supplies a partial explanation of the twofold diet of the Hymenoptera with carnivorous larvae: meat first, honey next. But then the note of interrogation is shifted. It stood elsewhere; it now stands here. Why is the Osmia, who as a larva fares so well on albumen, fed on honey at the start? Why do the Bee-tribe receive a vegetable diet when the other members of the order receive an animal diet?

If I were a believer in evolution, I should say yes, by the fact of its germ, every animal is originally carnivorous. The insect in particular starts with albuminoid materials. Many larvae adhere to the egg-food, many adult insects do likewise. But the struggle to fill the belly, which after all is the struggle for life, demands something better than the precarious hazards of the chase. Man, at first a ravenous hunter after game, brought the flock into existence and turned shepherd to avoid a time of dearth. An even greater progress inspired him to scrape the earth and to sow seed, which assures him of a living. The evolution from scarcity to moderation and from moderation to plenty has led to the resources of husbandry.

The animals forestalled us this path of progress. The ancestors of the Philanthus, in the remote ages of the lacustrian tertiary formations, lived by prey in both the larval and the adult forms: they hunted for themselves as well as for the family. They did not confine themselves to emptying the Bee's crop, as their descendants do to this day: they devoured the deceased. From the beginning to the end they remained flesh-eaters. Later, fortunate innovators, whose race supplanted the laggards, discovered an inexhaustible nourishment, obtained without dangerous conflicts or laborious search: the sugary secretions of the flowers. The costly habit of living on prey, which does not favour large populations, was maintained for the feeble larvae; but the vigorous adult broke herself of it to lead an easier and more prosperous life. Thus, gradually, was formed the Philanthus of our day; thus was acquired the twofold diet of the various predatory insects our contemporaries.

The Bee has done better still: from the moment of leaving the egg she delivered herself completely from foodstuffs the acquisition of which depended on chance. She discovered honey, the grubs' food. Renouncing the chase for ever and becoming an agriculturalist pure and simple, the insect attains a degree of physical and moral prosperity which the predatory species are far from sharing. Hence the flourishing colonies of the Anthophorae, the Osmiae, the Eucerae, the Halicti and other honey-manufacturers, whereas the predatory insects work in isolation; hence the societies in which the Bee displays her wonderful tendencies, the supreme expression of instinct.

This is what I should say if I belonged to that school. It all forms a chain of very logical deductions and proffers itself with a certain air of likelihood which we should be glad to find in a host of evolutionist arguments put forward as irrefutable. Well, I will make a present of my deductive views, without regret, to whoever cares to have them: I don't believe one word of them; and I confess my profound ignorance of the origin of the twofold diet.

What I do understand more clearly, after all these investigations, is the tactics of the Philanthus. When witnessing her ferocious feasting, the real reason of which was unknown to me, I heaped the most ill-sounding epithets upon her, calling her a murderess, a bandit, a pirate, a robber of the dead. Ignorance is always evil-tongued; the man who does not know indulges in rude assertions and mischievous interpretations. Now that my eyes have been opened to the facts, I hasten to apologize and to restore the Philanthus to her place in my esteem. In draining the crops of her Bees the mother is performing the most praiseworthy of all actions: she is protecting her family against poison. If she happens to kill on her own account and to abandon the corpse after making it disgorge, I dare not reckon this against her as a crime. When the habit has been formed of emptying the Bee's crop with a good motive, there is a great temptation to do it again with no other excuse than hunger. Besides, who knows? Perhaps there is always at the back of her hunting some thought of game which might be useful for the larvae. Although not carried into effect, the intention excuses the deed.

I therefore withdraw my epithets in order to admire the insect's maternal logic and to hold it up to the admiration of others. The honey would be pernicious to the health of the larvae. How does the mother know that the syrup, a treat for her, is unwholesome for her young? To this question our science offers no reply. The honey, I say, would imperil the grubs' lives, The Bee must therefore first be made to disgorge. The disgorging must be effected without lacerating the victim, which the nurseling must receive in the fresh state; and the operation is impracticable on a paralysed insect because of the resistance of the stomach. The Bee must therefore be killed outright instead of being paralysed, or the honey will not be voided. Instantaneous death can be inflicted only by wounding the primordial centre of life. The sting must therefore aim at the cervical ganglia, the seat of innervation on which the rest of the organism depends. To reach them there is only one way, through the little gap in the throat. It is here therefore that the sting must be inserted; and it is here in fact that it is inserted, in a spot hardly as large as the twenty-fifth of an inch square. Suppress a single link of this compact chain, and the Bee-fed Philanthus becomes impossible.

That honey is fatal to carnivorous larvae is a fact which teems with consequences. Several Hunting Wasps feed their families upon Bees. These include, to my knowledge, the Crowned Philanthus (P. coronatus, FAB.), who lines her burrows with big Halicti; the Robber Philanthus (P. raptor, LEP.), who chases all the smaller-sized Halicti, suited to her own dimensions, indifferently; the Ornate Cerceris (C. ornata, FAB.), another passionate lover of Halicti; and the Palarus (P. flavipes, FAB.), who, with a curious eclecticism, stacks in her cells the greater part of the Hymenopteron clan that does not exceed her powers. What do these four huntresses and the others of similar habits do with their victims whose crops are more or less swollen with honey? They must follow the example of the Bee-eating

Philanthus and make them disgorge, lest their family perish of a honeyed diet; they must manipulate the dead Bee, squeeze her and drain her dry. Everything goes to show it. I leave it to the future to display these dazzling proofs of my doctrine in their proper light.

CHAPTER 11.

THE METHOD OF THE AMMOPHILAE.

***

My readers may differ in appraising the comparative value of the trifling discoveries which entomology owes to my labours. The geologist, the recorder of forms, will prefer the hypermetamorphosis of the Oil-beetles, the development of the Anthrax or larval dimorphism; the embryogenist, searching into the mysteries of the egg, will have some esteem for my enquiries into the egg-laying habits of the Osmia; the philosopher, racking his brain over the nature of instinct, will award the palm to the operations of the Hunting Wasps. I agree with the philosopher. Without hesitation, I would abandon all the rest of my entomological baggage for this discovery, which happens to be the earliest in date and that of which I have the fondest memories. Nowhere do I find a more brilliant, more lucid, more eloquent proof of the intuitive wisdom of instinct; nowhere does the theory of evolution suffer a more obstinate check.

Darwin, a true judge, made no mistake about it. He greatly dreaded the problem of the instincts. My first results in particular left him very anxious. If he had known the tactics of the Hairy Ammophila, the Mantis-hunting Tachytes, the Bee-eating Philanthus, the Calicurgi and other marauders, his anxiety, I believe, would have ended in a frank admission that he was unable to squeeze instinct into the mould of his formula. Alas, the philosopher of Down quitted this world when the discussion, with experiments to support it, had barely begun: a method superior to any argument! The little that I had published at that time left him with still some hope of an explanation. In his eyes, instinct was always an acquired habit. The predatory Wasps killed their prey at first by stabbing it at random, here and there, in the softest parts. By degrees they found the spot where the sting was most effectual; and the habit once formed became a true instinct. Transitions from one method of operation to the other, intermediary changes, sufficed to bolster up these sweeping assertions. In a letter of the 16th of April, 1881, he asks G.J. Romanes to consider the problem:

"I do not know," he says "whether you will discuss in your book on the mind of animals any of the more complex and wonderful instincts. It is unsatisfactory work, as there can be no fossilised instincts, and the sole guide is their state in other members of the same order, and mere PROBABILITY."

But if you do discuss any (and it will perhaps be expected of you), I should think that you could not select a better case than that of the sand-wasps which paralyse their prey as described by Fabre in his wonderful paper in the "Anales des sciences naturelles," and since amplified in his admirable Souvenirs...

I thank you, O illustrious master, for your eulogistic expressions, proving the keen interest which you took in my studies of instinct, no ungrateful task - far from it - when we tackle it as it should be tackled: from the front, with the aid of facts, and not from the flank, with the aid of arguments. Arguments are here out of place, if we wish to maintain our position in the light. Besides, where would they lead us? To evoking the instincts of bygone ages, which have not been preserved by fossilization? Any such appeal to the dim and distant past is quite unnecessary, if we wish for variations of instinct, leading by degrees, according to you, from one instinct to another; the present world offers us plenty.

Each operator has her particular method, her particular kind of game, her particular points of attack and tricks of fence; but in the midst of this variety of talents we observe, immutable and predominant, the perfect accordance of the surgery with the victim's organization and the larva's needs. The art of one will not explain the art of another, no less exact in the delicacy of its rules. Each operator has her own tactics, which tolerate no apprenticeship. The Ammophila, the Scolia, the Philanthus and the others all tell us the same thing: none can leave descendants if she be not from the outset the skilful paralyser or slayer that she is today. The "almost" is impracticable when the future of the race is at stake. What would have become of the firstborn mammal but for its perfect instinct of suckling?

And then, to suppose the impossible: a Wasp discovers by chance the operative method which will be the saving attribute of her race. How are we to admit that this fortuitous act, to which the mother has vouchsafed no more attention than to her other less fortunate attempts, could leave a profound trace behind it and be faithfully transmitted by heredity? Is it not going beyond reason, going beyond the little that is known to us as certain, if we grant to atavism this strange power, of which our present world knows no instance? There is a good deal to be said for this point of view, my revered master! But, once more, arguments are here out of place; there is room only for facts, of which I will resume the recital.

Hitherto I had but one means of studying the operative methods of the spoilers: to surprise the Wasp in possession of her capture, to rob her of her prey and immediately to give her in exchange a similar prey, but a living one. This method of substitution is an excellent expedient. Its only defect - a very grave one - is that it subjects observation to very uncertain chances. There is little prospect of meeting the insect dragging its victim along; and, in the second place, should good fortune suddenly smile upon you, preoccupied as you are with other matters you have not the substitute at hand. If we provide ourselves with the necessary head of game in advance, the huntress is not there. We avoid one reef to founder on another. Moreover, these unlooked for observations, made sometimes on the public highway, the worst of laboratories, are only half-satisfactory. In the case of swiftly-enacted scenes, which it is not in our power to renew again and again until perfect conviction is reached, we always fear lest we may not have seen accurately, may not have seen everything.

A method which could be controlled at will would offer the best guarantees, above all if employed at home, under comfortable conditions, favourable to precision. I wished, therefore, to see my insects at work on the actual table at which I am writing their history. Here very few of their secrets would escape me. This wish of mine was an old one. As a beginner, I made some experiments under glass with the Great Cerceris (C. tuberculata) and the Yellow-winged Sphex. Neither of them responded to my desires. The refusal of each to attack respectively her Cleonus or her Cricket discouraged further progress in this direction. I was wrong to abandon my attempts so soon. Now, very long afterwards, the idea occurs to me to place under glass the

Bee-eating Philanthus, whom I sometimes surprise in the open engaged in forcing a bee to disgorge her honey. The captive massacres her bees in such a spirited fashion that the old hope revives stronger than ever. I contemplate reviewing all the wielders of the stiletto and forcing each to reveal her tactics.

I was obliged to abate these ambitions considerably. I had some successes and many more failures. I will tell you of the former. My insect-cage is a spacious dome of wire-gauze resting on a bed of sand. Here I keep in reserve the captives of my hunting-expeditions. I feed them on honey, placed in little drops on spikes of lavender, on heads of thistle, or field eryngo, or globe-thistle, according to the season. Most of my prisoners do well on this diet and seem scarcely affected by their internment; others pine away and die in two or three days. These victims of despair nearly always throw me back, because of the difficulty of obtaining the necessary prey at short notice.

Indeed it entails no small trouble to secure in the nick of time the game demanded by the huntress who has recently fallen a captive to my net. As assistant-purveyors I have a few small schoolboys, who, released from the tedium of their declensions and conjugations, set out, on leaving the classroom, to inspect the greenswards and beat the bushes in the neighbourhood on my behalf. The gros sou, the penny-piece, if you please, stimulates their zeal; but with misadventurous results! What I need today is Crickets. The band sallies forth and returns with not a single Cricket, but numbers of Ephippigers, for which I asked the day before yesterday and which I no longer need, my Languedocian Sphex being dead. General surprise at this sudden change of market. My young scatterbrains find it hard to understand that the beast which was so precious two days ago is now of no value whatever. When, owing to the chances of my net, a renewed demand for the Ephippiger sets in, then they will bring me the Cricket, the despised Cricket.

Such a trade could never hold out if now and again my speculators were not encouraged by some success. At the moment when urgent necessity is sending up prices, one of them brings me a magnificent gadfly intended for the Bembex. For two hours, when the sun was at its height, he kept watch on the threshing-floor hard by, waiting for the bloodsucker, in order to catch him on the buttocks of the Mules which trot round and round trampling the corn. This gallant fellow shall have his gros sou and a slice of bread and jam as well. A second, no less fortunate, has found a fat Spider, the Epeira, for whom my Pompili are waiting. To the two sous of this fortunate youth I add a little picture for his missal. Thus are my purveyors kept going; and, after all, their help would be very inadequate if I did not take upon myself the main burden of these wearisome quests.

Once in possession of the requisite prey, I transfer the huntress from my warehouse, the wire-gauze cage, to a bell-glass varying in capacity from one to three or four litres, according to the size and habits of the combatants; I place the victim in the arena; I expose the bell-glass to the direct rays of the sun, without which condition the executioner as a rule declines to operate; I arm myself with patience and await events.

We will begin with the Hairy Ammophila, my neighbour. Year after year, when April comes, I see her in considerable numbers, very busy on the paths in my enclosure.

Until June I see her digging her burrows and searching for the Grey Worm, to be placed in the meat-cellar. Her tactics are the most complex that I know and more than any other deserves to be thoroughly studied. To capture the cunning vivisector, to release her and catch her again I find an easy matter for the best part of a month; she works outside my door.

I have still to obtain the Grey Worm. This means a repetition of the disappointments which I had before, when, to find a caterpillar, I was obliged to watch the Ammophila while hunting and to be guided by her hints, as the truffle-hunter is guided by the scent of his Dog. A patient exploration of the harmas, one tuft of thyme after another, does not give me a single worm. My rivals in this search are finding their game at every moment; I cannot find it even once. Yet one more reason for bowing to the superiority of the insect in the management of her affairs. My band of schoolboys get to work in the surrounding fields. Nothing, always nothing! I in my turn explore the outer world; and for ten days the pursuit of a caterpillar torments me till I lose my power of sleep. Then, at last, victory! At the foot of a sunny wall, under the budding rosettes of the panicled centaury, I find a fair supply of the precious Grey Worm or its equivalent.

Behold the worm and the Ammophila face to face beneath the bell-glass. Usually the attack is prompt enough. The caterpillar is grabbed by the neck with the mandibles, wide, curved pincers capable of embracing the greater part of the living cylinder. The creature thus seized twists and turns and sometimes, with a blow of its tail, sends the assailant rolling to a distance. The latter is unconcerned and thrusts her sting thrice in rapid succession into the thorax, beginning with the third segment and ending with the first, where the weapon is driven home with greater determination than elsewhere.

The caterpillar is then released. The Ammophila stamps on the ground; with her quivering tarsi she taps the cardboard on which the bell-glass stands; she lies down flat, drags herself along, gets up again, flattens herself once more. The wings jerk convulsively. From time to time the insect places its mandibles and forehead on the ground, then rears high upon its hind-legs as though to turn head over heels. In all this I see a manifestation of delight. We rub our hands when rejoicing at a success; the Ammophila is celebrating her triumph over the monster in her own fashion. During this fit of delirious joy, what is the wounded caterpillar doing? It can no longer walk; but all the part behind the thorax struggles violently, curling and uncurling when the Ammophila sets a foot upon it. The mandibles open and shut menacingly.

SECOND ACT. - When the operation is resumed, the caterpillar is seized by the back. From front to rear, in order, all the segments are stung on the ventral surface, except the three operated on. All serious danger is averted by the stabs of the first act; therefore, the Wasp is now able to work upon her patient without the haste displayed at the outset. Deliberately and methodically she drives in her lancet, withdraws it, selects the spot, stabs it and begins again, passing from segment to segment, taking care, each time, to lay hold of the back a little more to the rear, in order to bring the segment to be paralysed within reach of the needle. For the second time, the caterpillar is released. It is absolutely inert, except the mandibles, which are still capable of biting.

THIRD ACT. - The Ammophila clasps the paralysed victim between her legs; with the hooks of her mandibles she seizes the back of its neck, at the base of the first thoracic segment. For nearly ten minutes she munches this weak spot, which lies close to the cerebral nerve-centres. The pincers squeeze suddenly but at intervals and methodically, as though the manipulator wished each time to judge of the effect produced; the squeezes are repeated until I am tired of trying to count them. When they cease, the caterpillar's mandibles are motionless. Then comes the transportation of the carcase, a detail which is not relevant in this place.

I have set forth the complete tragedy, as it is fairly often enacted, but not always. The insect is not a machine, unvarying in the effect of its mechanism; it is allowed a certain latitude, enabling it to cope with the eventualities of the moment. Any one expecting to see the incidents of the struggle unfolding themselves exactly as I have described will risk disappointment. Special instances occur - they are even numerous - which are more or less at variance with the general rule. It will be well to mention the more important, in order to put future observers on their guard.

Not infrequently the first act, that of paralysing the thorax, is restricted to two thrusts of the sting instead of three, or even to one, which is then delivered in the foremost segment. This, it would seem, from the persistency with which the Ammophila inflicts it, is the most important prick of all. Is it unreasonable to suppose that the operator, when she begins by pricking the thorax, intends to subdue her capture and to make it incapable of injuring her, or even of disturbing her when the moment comes for the delicate and protracted surgery of the second act? This idea seems to me highly admissible; and then, instead of three dagger-thrusts, why not two only, why not merely one, if this would suffice for the time being? The amount of vigour displayed by the caterpillar must be taken into consideration. Be this as it may, the segments spared in the first act are stabbed in the second. I have sometimes even seen the three thoracic segments stung twice over: at the beginning of the attack and again when the Wasp returned to her vanquished prey.

The Ammophila's triumphant transports beside her wounded and writhing victim are also subject to exceptions. Sometimes, without releasing its prey for a moment, the insect proceeds from the thorax to the next segments and completes its operation in a single spell. The joyous entr'acte does not take place; the convulsive movements of the wings and the acrobatic postures are suppressed.

The rule is paralysis of all the segments, however many, in regular order from front to back, including even the anal segment if this boast of legs. By a fairly frequent exception the last two or three segments are spared. Another exception, but a very rare one, of which I have observed only a single instance, consists in the inversion of the dagger-thrusts of the second act, the thrusts being delivered from back to front. The caterpillar is then seized by its hinder extremity; and the Ammophila, progressing towards the head, stings in reverse order, passing from the succeeding to the preceding segment, including the thorax already stabbed. This reversal of the usual tactics I am inclined to attribute to negligence on the insect's part. Negligence or not, the inverted method has the same final result as the direct method: the paralysis of all the segments.

Lastly, the compression of the neck by the mandibulary pincers, the munching of the weak spot between the base of the skull and the first segment of the thorax, is sometimes practised and sometimes neglected. If the caterpillar's jaws open and threaten, the Ammophila stills them by biting the neck; if they are already growing quiescent, she refrains. Without being indispensable, this operation is useful at the moment of carting the prey. The caterpillar, too heavy to be carried on the wing, is dragged, head first, between the Ammophila's legs. If the mandibles are working, the least clumsiness may render them dangerous to the carrier, who is exposed to their bite without any means of defence.

Moreover, once on the way, thickets of grass are traversed in which the Grey Worm can seize a blade and offer a desperate resistance to the traction. Nor is this all. The Ammophila does not as a rule trouble about her burrow, or at least does not complete it, until she has caught her caterpillar. During the mining-operations, the game is laid somewhere high up, out of reach of the Ants, on some tuft of grass, or the twigs of a shrub, whither the huntress, from time to time, stopping her well-sinking, hastens to see if her quarry is still there. For her this is a means of refreshing her memory of the spot where she has laid it, often at some distance from the burrow, and of preventing attempts at robbery. When the moment comes for removing the game from its hiding-place, the difficulty would be insurmountable were the worm, gripping the shrub with all the might of its jaws, to anchor itself there. Hence inertia of the powerful hooks, which are the paralysed creature's sole means of resistance, becomes essential during the carting. The Ammophila obtains it by compressing the cerebral ganglia, by munching the neck. The inertia is temporary; it wears off sooner or later; but by this time the carcase is in the cell and the egg, prudently laid at a distance on the ventral surface of the worm, has nothing to fear from the caterpillar's grapnels. No comparison is permissible between the methodical squeezes of the Ammophila benumbing the cephalic nerve-centres and the brutal manipulations of the Philanthus emptying the crop of her Bee. The huntress of Grey Worms induces a temporary torpor of the mandibles; the ravisher of Bees makes them eject their honey. No one gifted with the least perspicacity will confound the two operations.

For the moment we will not dwell any longer on the method of the Hairy Ammophila; we will see instead how her kinswomen behave. After protracted refusals the Sandy Ammophila (A. sabulosa, FAB.),on whom I experimented in September, ended by accepting the proffered prey, a powerful caterpillar as thick as a lead-pencil. The surgical method did not differ from that employed by the Hairy Ammophila when operating on her Grey Worm in one spell. All the segments, excepting the last three, were stung from front to back, beginning with the prothorax. This single success with a simplified method left me in ignorance of the accessory manoeuvres, which I do not doubt must more or less closely recall those of the preceding species.

I am all the more inclined to accept these secondary manoeuvres, not as yet recorded - the transports of triumph and the compressions of the neck - inasmuch as I see them practised upon the Looper caterpillars, which differ so greatly from the others in external structure, exactly as I have described them in the case of the Grey Worm, which is of the ordinary form. Two species, the Silky Ammophila (A. holoserica, FAB.) and Jules' Ammophila,[3] affect this curious prey, which moves with the stride of a pair of compasses. The first, often renewed under glass during the greater part of August, has always refused my offers; the second, her contemporary, has, on the contrary, promptly accepted them.

I present Jules' Ammophila with a slender, brownish Looper which I caught on the jasmine. The attack is not slow in coming. The caterpillar is grabbed by the neck: lively contortions of the victim, which rolls the aggressor over and drags her along, now uppermost, now undermost in the struggle. First the thorax is stung, in its three rings, from back to front. The sting lingers longest near the throat, in the first segment. This done, the Ammophila releases her victim and proceeds to stamp her tarsi, to polish her wings, to stretch herself. Again I observe the acrobatic postures, the forehead touching the ground, the hinder part of the body raised. This mimic triumph is the same as that of the huntress of the Grey Worm. Then the Looper is once more seized. Despite its contortions, which are not in the least abated by the three wounds in the thorax, it is stung from front to back in each segment still unwounded, no matter how many, whether supplied with legs or not. I expected to see the sting refrain more or less in the long interval which separates the true legs in front from the pro-legs at the back: segments devoid of organs of defence or locomotion did not seem to me to deserve conscientious

3 See in the first volume of the "Souvenirs entomologiques" what I mean by this denomination.

surgery. I was mistaken: not a segment of the Looper is spared, not even the last ones. It is true that these, being eminently capable of catching hold with their false legs, would be dangerous later were the Wasp to neglect them.

I observe, however, that the lancet works more rapidly in the second part of the operation than in the first, either because the caterpillar, half subjugated by the triple wound at the outset, is easier to reach with the sting, or because the segments more remote from the head are rendered harmless with a smaller injection of poison. Nowhere do we see repeated the care expended upon paralysing the thorax, still less the insistent attention to the first segment. On returning to her Looper after the entr'acte devoted to the joys of success, the Ammophila stabs so swiftly that, on one occasion, I saw her obliged to begin all over again. Lightly stung along its whole length, the victim still struggles. Without hesitation, the operator unsheathes her scalpel for the second time and operates on the Looper afresh, with the exception of the thorax, which was already sufficiently anaesthetized. This done, all is in order; there is no more movement.

After the stiletto the hooks of the mandibles rarely fail to intervene. Long and curved, they nibble at the paralysed victim's neck, sometimes from above, sometimes from below. It is a repetition of what the Hairy Ammophila showed us: the same sudden squeezes of the pincers, with rather long intervals between. These intervals, these measured bites and the insect's watchful attitude have every appearance of telling us that the operator is noting the effect produced before giving a fresh pinch of the nippers.

It will be seen how valuable is the evidence of Jules' Ammophila: it tells us that the immolaters of Looper caterpillars and those of ordinary caterpillars follow precisely the same method; that victims displaying very dissimilar external structure do not entail any modification of the operative tactics so long as the internal organization remains the same. The number, arrangement and degree of mutual independence of the nerve-centres guide the sting; the anatomy of the game, rather than its form, controls the huntress' tactics.

Let me mention, before I dismiss the subject, a superb example of this marvellous anatomical discrimination. I once took from between the legs of a Hairy Ammophila, which had just paralysed it, a caterpillar of Dicranura vinula. What a strange capture compared with the ordinary caterpillar! Bridling in thick folds beneath its pink neckerchief, its forepart raised in a sphinx-like attitude, its hinder-part slowly waving two long caudal threads, the curious animal is no caterpillar to the schoolboy who brings it to me, nor to the man who comes upon it while cutting his bundle of osiers; but it is a caterpillar to the Ammophila, who treats it accordingly. I explore the queer creature's segments with the point of a needle. All are insensitive; all therefore have been stung.

CHAPTER 12.

THE METHOD OF THE SCOLIAE.

***

After the Ammophilae, the paralysers who multiply their lancet-thrusts to destroy the influence of the various nerve-centres, excepting those of the head, it seemed advisable to interrogate other insects which also are accustomed to a naked prey, vulnerable at all points save the head, but which deliver only a single thrust of the sting. Of these two conditions the Scoliae fulfilled one, with their regular quarry, the tender Cetonia-, Oryctes- or Anoxia-larva, according to the Scolia's species. Did they fulfil the second? I was convinced beforehand that they did. From the anatomy of the victims, with their concentrated nervous system, I foresaw, when compiling my history of the Scoliae, that the sting would be unsheathed once only; I even mentioned the exact spot into which the weapon would be plunged.

These were assertions dictated by the anatomist's scalpel, without the slightest direct proof derived from observed facts. Manoeuvres executed underground escaped the eye, as it seemed to me that they must always do. How indeed could I hope that a creature whose art is practised in the darkness of a heap of mould would decide to work in broad daylight? I did not reckon upon it all. Nevertheless, to salve my conscience, I tried bringing the Scolia into contact with her prey under the bell-glass. I was well-advised to do so, for my success was in inverse ratio to my hopes. Next to the Philanthus, none of the Hunting Wasps displayed such ardour in attacking under artificial conditions. All the insects experimented upon, some sooner, some later, rewarded me for my patience. Let us watch the Two-banded Scolia (S. bifasciata, VAN DER LIND) operating on her Cetonia grub.

The incarcerated larva strives to escape its terrible neighbour. Lying on its back, it fiercely wends its way round and round the glass circus. Presently the Scolia's attention awakens and is betrayed by a continued tapping with the tips of the antennae upon the table, which now represents the accustomed soil. The Wasp attacks the game, delivering her assault upon the monster's hinder end. She climbs upon the Cetonia-grub, obtaining a purchase with the tip of her abdomen. The quarry merely travels the more quickly on its back, without coiling itself into a defensive posture. The Scolia reaches the forepart, with tumbles and other accidents which vary greatly with the amount of tolerance displayed by the larva, her improvised steed. With her mandibles she nips a point of the thorax, on the upper surface; she places herself athwart the beast, arches herself and makes every effort to reach with the end of her abdomen the region into which the sting is to be driven. The arch is a little too narrow to embrace almost the whole circumference of her corpulent prey; and she renews her attempts and efforts for a long time. The tip of the belly tries every conceivable expedient, touching here, there and everywhere, but as yet stopping nowhere. This persistent search in itself demonstrates the importance which the paralyser attaches to the point at which her lancet is to penetrate the flesh.

Meanwhile, the larva continues to move along on its back. Suddenly it curls up; with a stroke of its head it hurls the enemy to a distance. Undiscouraged by all her setbacks, the Wasp picks herself up, brushes her wings and resumes her attack upon the colossus, almost always by mounting the larva's hinder end. At last after all these fruitless attempts, the Scolia succeeds in achieving the correct position. She is seated athwart the Cetonia-grub; the mandibles grip a point on the dorsal surface of the thorax; the body, bent into a bow, passes under the larva and with the tip of the belly reaches the region of the neck. The Cetonia-grub, placed in serious peril, writhes, coils and uncoils itself, spinning round upon its axis. The Scolia does not interfere. Holding the victim tightly gripped, she turns with it, allows herself to be dragged upwards, downwards, sidewards, following its contortions. Her obstinacy is such that I can now remove the bell-glass and follow the details of the drama in the open.

Briefly, in spite of the turmoil, the tip of the abdomen feels that the right spot has been found. Then and only then the sting is unsheathed. It plunges in. The thing is done. The larva, at first plump and active, suddenly becomes flaccid and inert. It is paralysed. Henceforth there are no movements save of the antennae and the mouthparts, which will for a long time yet bear witness to a remnant of life. The point wounded has never varied in the series of combats under glass: it occupies the middle of the line of demarcation between the prothorax and the mesothorax, on the ventral surface. Note that the Cerceres, operating on Weevils, whose nervous system is as compact as the Cetonia-grub's, drive in the needle at the same spot. Similarity of nervous organization occasions similarity of method. Note also that the Scolia's sting remains in the wound for some time and roots about with marked persistence. Judging by the movements of the tip of the abdomen, one would certainly say that the weapon is exploring and selecting.

Free to shift in one direction or the other, within narrow limits, its point is most probably seeking for the little mass of nerve-tissue which must be pricked, or at least sprinkled with poison, to obtain overwhelming paralysis.

I will not close this report of the duel without relating a few further facts, of minor importance. The Two-banded Scolia is a fierce persecutor of the Cetonia. In one sitting the same mother stabs three larvae, one after the other, in front of my eyes. She refuses the fourth, perhaps owing to fatigue or to exhaustion of the poison-bag. Her refusal is only temporary. Next day, she begins again and paralyses two grubs; the day after that, she does the same, but with a zeal that decreases from day to day.

The other Hunting Wasps that pursue the chase far afield grip, drag, carry their prey, after depriving it of movement, each in her own fashion and, laden with their burden, make prolonged attempts to escape from the bell-glass and to gain the burrow. Discouraged by these futile endeavours, they abandon them at last. The Scolia does not remove her quarry, which lies on its back for an indefinite time on the actual spot of the sacrifice. When she has withdrawn her dagger from the wound, she leaves her victim where it lies and, without taking further notice of it, begins to flutter against the side of the glass. The paralysed carcase is not transported elsewhere, into a special cellar; there where the struggle has occurred it receives, upon its extended abdomen, the egg whence the consumer of the succulent tit-bit will emerge, thus saving the expense of setting up house. It goes without saying that under the bell-glass the laying does not take place: the mother is too cautious to abandon her egg to the perils of the open air.

Why then, recognizing the absence of her underground burrow, does the Scolia uselessly pursue the Cetonia with the frantic ardour of the Philanthus flinging herself upon the Bee? The action of the Philanthus is explained by her passion for honey; hence the murders committed in excess of the needs of her family. The Scolia leaves us perplexed: she takes nothing from the Cetonia-grub, which is left without an egg; she stabs, though well aware of the uselessness of her action: the heap of mould is lacking and it is not her custom to transport her prey. The other prisoners, once the blow is struck, at least seek to escape with their capture between their legs; the Scolia attempts nothing.

After due reflection, I lump together in my suspicions all these surgeons and ask myself whether they possess the slightest foresight, where the egg is concerned. When, exhausted by their burden, they recognize the impossibility of escape, the more expert among them ought not to begin all over again; yet they do so begin a few minutes later. These wonderful anatomists know absolutely nothing about anything, they do not even know what their victims are good for. Admirable artists in killing and paralysis, they kill or paralyse at every favourable opportunity, no matter what the final result as regards the egg. Their talent, which leaves our science speechless, has not a shadow of consciousness of the task accomplished.

A second detail strikes me: the desperate persistence of the Scolia. I have seen the struggle continue for more than a quarter of an hour, with frequent alternations of good luck and bad, before the Wasp achieved the required position and reached with the end of her abdomen the spot where the sting should penetrate.

During these assaults, which were resumed as soon as they were repulsed, the aggressor repeatedly applied the tip of her belly to the larva, but without unsheathing, as I could see by the absence of the start which the larva gives when it feels the pain of the sting. The Scolia therefore does not prick the Cetonia anywhere until the weapon covers the requisite spot. If no wounds are inflicted elsewhere, this is not in any way due to the structure of the larva, which is soft and vulnerable all over, except in the head. The point sought by the sting is no more unprotected than any other part of the skin.

In the scuffle, the Scolia, curved into a bow, is sometimes seized by the vice-like grip of the Cetonia-grub, which is violently coiling and uncoiling. Heedless of the powerful grip, the Wasp does not let go for a moment, either with her mandibles or with the tip of her abdomen. At such times the two creatures, locked in a mutual embrace, turn over and over in a mad whirl, each of them now on top, now underneath. When it contrives to rid itself of its enemy, the larva uncoils again, stretches itself out and proceeds to make off upon its back with all possible speed. Its defensive ruses are exhausted. Formerly, before I had seen things for myself, taking probability as my guide I willingly granted to the larva the trick of the Hedgehog, who rolls himself into a ball and sets the Dog at defiance. Coiled upon itself, with an energy which my fingers have some difficulty in overcoming, the larva, I thought, would defy the Scolia, powerless to unroll it and disdaining any point but the one selected. I hoped and believed that it possessed this means of defence, a means both efficacious and extremely simple. I had presumed too much upon its ingenuity. Instead of imitating the Hedgehog and remaining contracted, it flees, belly in air; it foolishly adopts the very posture which allows the Scolia to mount to the assault and to reach the spot for the fatal stroke. The silly beast reminds me of the giddy Bee who comes and flings herself into the clutches of the Philanthus. Yet another who has learnt no lesson from the struggle for life.

Let us proceed to further examples. I have just captured an Interrupted Scolia (Colpa interrupta, LATR.), exploring the sand, doubtless in search of game. It is a matter of making the earliest possible use of her, before her spirit is chilled by the tedium of captivity. I know her prey, the larva of Anoxia australis; I know, from my past excavations, the points favoured by the grub: the mounds of sand heaped up by the wind at the foot of the rosemaries on the neighbouring hillsides. It will be a hard job to find it, for nothing is rarer than the common if one wants it then and there. I appeal for assistance to my father, an old man of ninety, still straight as a capital I. Under a sun hot enough to broil an egg, we set off, shouldering a navvy's shovel and a three-pronged luchet. Employing our feeble energies in turns, we dig a trench in the sand where I hope to find the Anoxia. My hopes are not disappointed. After having by the sweat of our brow - never was the expression more justified - removed and sifted two cubic yards at least of sandy soil with our fingers, we find ourselves in possession of two larvae. If I had not wanted any, I should have turned them up by the handful. But my poor and costly harvest is sufficient for the moment. tomorrow I will send more vigorous arms to continue the work of excavation.

And now let us reward ourselves for our trouble by studying the tragedy in the bell-glass. Clumsy, awkward in her movements, the Scolia slowly goes the round of the circus. At the sight of the game, her attention is aroused. The struggle is announced by the same preparations as those displayed by the Two-banded Scolia: the Wasp polishes her wings and taps the table with the tips of her antennae. And view, halloo! The attack begins. Unable to move on a flat surface, because of its short and feeble legs, deprived moreover of the Cetonia-larva's eccentric means of travelling on its back, the portly grub has no thought of fleeing; it coils itself up. The Scolia, with her powerful pincers, grips its skin now here, now elsewhere. Curved into a circle with the two ends almost touching, she strives to thrust the tip of her abdomen into the narrow opening in the coil formed by the larva. The contest is conducted calmly, without violent bouts at each varying accident. It is the determined attempt of a living split ring trying to slip one of its ends into another living split ring, which with equal determination refuses to open. The Scolia holds the victim subdued with her legs and mandibles; she tries one side, then the other, without managing to unroll the circle, which contracts still more as it feels its danger increasing. The actual circumstances make the operation more difficult: the prey slips and rolls about the table when the insect handles it too violently; there are no points of purchase and the sting cannot reach the desired spot; the fruitless efforts are continued for more than an hour, interrupted by periods of rest, during which the two adversaries represent two narrow, interlocked rings.

What ought the powerful Cetonia-grub to do to defy the Two-banded Scolia, who is far less vigorous than her victim? It should imitate the Anoxia-larva and remain rolled up like a Hedgehog until the enemy retires. It tries to escape, unrolls itself and is lost. The other does not stir from its posture of defence and resists successfully.

Is this due to acquired caution? No, but to the impossibility of doing otherwise on the slippery surface of a table. Clumsy, obese, weak in the legs, curved into a hook like the common White Worm, the Anoxia-larva is unable to move along a smooth surface; it writhes laboriously, lying on its side. It needs the shifting soil in which, using its mandibles as a ploughshare, it digs into the ground and buries itself.

Let us try if sand will shorten the struggle, for I see no end to it yet, after more than an hour of waiting. I lightly powder the arena. The attack is resumed with a vengeance. The larva, feeling the sand, its native element, tries to escape. Imprudent creature! Did I not say that its obstinacy in remaining rolled up was due to no acquired prudence but to the necessity of the moment? The sad experience of past adversities has not yet taught it the precious advantage which it might derive from keeping its coils closed so long as danger remains. For that matter, on the unyielding support of my table, they are not one and all so cautious. The larger seem even to have forgotten what they knew so well in their youth: the defensive art of coiling themselves up.

I continue my story with a fine-sized specimen, less likely to slip under the Scolia's onslaught. When attacked, the larva does not curl up, does not shrink into a ring as did the last, which was younger and only half as large. It struggles awkwardly, lying on its side, half-open. For all defence it twists about; it opens, closes and reopens the great hooks of its mandibles. The Scolia grabs it at random, clasps it in her shaggy legs and for nearly a quarter of an hour battles with the luscious tit-bit. At last, after a not very tumultuous struggle, when the favourable position is attained and the propitious moment has come, the sting is implanted in the creature's thorax, in a central point, below the throat, level with the forelegs. The effect is instantaneous: total inertia, except of the appendages of the head, the antennae and mouthparts. I achieved the same results, the same prick at a definite, invariable point, with my several operators, renewed from time to time by some lucky cast of the net.

Let us mention, in conclusion, that the attack of the Interrupted Scolia is far less fierce than that of the Two-banded Scolia. The Wasp, a rough sand-digger, has a clumsy gait; her movements are stiff and almost automatic. She does not find it easy to repeat her dagger-thrust. Most of the specimens with which I experimented refused a second victim on the first two days after their exploits. As though somnolent, they did not stir unless excited by my teasing them with a bit of straw. Although more active and more ardent in the chase, the Two-banded Scolia likewise does not draw her weapon every time that I invite her. For all these huntresses there are moments of inaction which the presence of a fresh prey is powerless to disturb.

The Scoliae have taught me nothing further, in the absence of subjects belonging to other species. No matter: the results obtained represent no small triumph for my ideas. Before seeing the Scoliae operate, I said, guided solely by the anatomy of the victims, that the Cetonia-, Anoxia – and Oryctes-larvae must be paralysed by a single thrust of the lancet; I even named the point where the sting must strike, a central point, in the immediate vicinity of the forelegs. Of the three genera of paralysers, two have allowed me to witness their surgical methods, which the third, I feel certain, will confirm. In both cases, a single thrust of the lancet; in both cases, injection of the venom at a predetermined point. A calculator in an observatory could not compute the position of his planet with greater accuracy. An idea may be taken as proved when it attains to this mathematical forecast of the future, this certain knowledge of the unknown. When will the acclaimers of chance achieve a like success? Order appeals to order; and chance knows no laws.

CHAPTER 13.

THE METHOD OF THE CALICURGI.

***

The non-armoured victims, vulnerable by the sting over almost their whole body, ordinary caterpillars and Looper caterpillars, Cetonia – and Anoxia-larvae, whose only means of defence, apart from their mandibles, consists of rollings and contortions, called for the testimony of another victim, the Spider, almost as ill-protected, but armed with formidable poison-fangs. How, in particular, will the Ringed Calicurgus set to work in operating on the Black-bellied Tarantula, the terrible Lycosa, who with a single bite kills the Mole or the Sparrow and endangers the life of man? How does the bold Pompilus overcome an adversary more powerful than herself, better-equipped with virulent poison and capable of making a meal of her assailant? Of all the Hunting Wasps, none risks such unequal conflicts, in which appearances would proclaim the aggressor to be the victim and the victim the aggressor.

The problem was one deserving patient study. True, I foresaw, from the Spider's organization, a single sting in the centre of the thorax; but that did not explain the victory of the Wasp, emerging safe and sound from her tussle with such a quarry. I had to see what occurred. The chief difficulty was the scarcity of the Calicurgus. It is easy for me to obtain the Tarantula at the desired moment: the part of the plateau in my neighbourhood left untilled by the vine-growers provides me with as many as are necessary. To capture the Pompilus is another matter. I have so little hope of finding her that special quests are regarded as useless. To search for her would perhaps be just the way not to find her. Let us rely on the uncertainties of chance. Shall I get her or shall I not?

I've got her. I catch her unexpectedly on the flowers. Next day I supply myself with half a dozen Tarantulae. Perhaps I shall be able to employ them one after the other in repeated duels. As I return from my Lycosa-hunt, luck smiles upon me again and crowns my desires. A second Calicurgus offers herself to my net; she is dragging her heavy, paralysed Spider by one leg, in the dust of the highway. I attach great value to my find: the laying of the egg has become a pressing matter; and the mother, I believe, will accept a substitute for her victim without much hesitation. Here then are my two captives, each under her bell-glass with her Tarantula.

I am all eyes. What a tragedy there will be in a moment! I wait, anxiously . . . But . . . but . . . what is this? Which of the two is the assailed? Which is the assailant? The characters seem to be inverted. The Calicurgus, unable to climb up the smooth glass wall, strides round the ring of the circus. With a proud and rapid gait, her wings and antennae vibrating, she goes and returns. The Lycosa is soon seen. The Calicurgus approaches her without the least sign of fear, walks round her and appears to have the intention of seizing one of her legs. But at that moment the Tarantula rises almost vertically on her four hinder legs, with her four front legs lifted and outspread, ready for the counterstroke. The poison-fangs gape widely; a drop of venom moistens their tips. The very sight of them makes my flesh creep. In this terrible attitude, presenting her powerful thorax and the black velvet of her belly to the enemy, the Spider overawes the Pompilus, who suddenly turns tail and moves away. The Lycosa then closes her bundle of poisoned daggers and resumes her natural pose, standing on her eight legs; but, at the slightest attempt at aggression on the Wasp's part, she resumes her threatening position.

She does more: suddenly she leaps and flings herself upon the Calicurgus; swiftly she clasps her and nibbles at her with her fangs. Without wielding her sting in self-defence, the other disengages herself and merges unscathed from the angry encounter. Several times in succession I witness the attack; and nothing serious ever befalls the Wasp, who swiftly withdraws from the fray and appears to have received no hurt. She resumes her marching and countermarching no less boldly and swiftly than before.

Is this Wasp invulnerable, that she thus escapes from the terrible fangs? Evidently not. A real bite would be fatal to her. Big, sturdily built Acridians succumb; how is it that she, with her delicate organism, does not! The Spider's daggers, therefore, make no more than an idle feint; their points do not enter the flesh of the tight-clasped Wasp. If the strokes were real, I should see bleeding wounds, I should see the fangs close for a moment on the part seized; and with all my attention I cannot detect anything of the kind. Then are the fangs powerless to pierce the Wasp's integuments? Not so. I have seen them penetrate, with a crackling of broken armour, the corselet of the Acridians, which offers a far greater resistance. Once again, whence comes this strange immunity of the Calicurgus held between the legs and assailed by the daggers of the Tarantula? I do not know. Though in mortal peril from the enemy confronting her, the Lycosa threatens her with her fangs and cannot decide to bite, owing to a repugnance which I do not undertake to explain.

Obtaining nothing more than alarums and excursions of no great seriousness, I think of modifying the gladiatorial arena and approximating it to natural conditions. The soil is very imperfectly represented by my worktable; and the Spider has not her fortress, her burrow, which plays a part of some importance both in attack and in defence. A short length of reed is planted perpendicularly in a large earthenware pan filled with sand. This will be the Lycosa's burrow. In the middle I stick some heads of globe-thistle garnished with honey as a refectory for the Pompilus; a couple of Locusts, renewed as and when consumed, will sustain the Tarantula. These comfortable quarters, exposed to the sun, receive the two captives under a wire-gauze dome, which provides adequate ventilation for a prolonged residence.

My artifices come to nothing; the session closes without result. A day passes, two days, three; still nothing happens. The Pompilus is assiduous in her visits to the honeyed flower-clusters; when she has eaten her fill, she clambers up the dome and makes interminable circuits of the netting; the Tarantula quietly munches her Locust. If the other passes within reach, she swiftly raises herself and waves her off. The artificial burrow, the reed-stump, fulfills its purpose excellently. The Lycosa and the Pompilus resort to it in turns, but without quarrelling. And that is all. The drama whose prologue was so full of promise appears to be indefinitely postponed.

I have a last resource, on which I base great hopes: it is to remove my two Calicurgi to the very site of their investigations and to install them at the door of the Spider's lodging, at the top of the natural burrow. I take the field with an equipment which I am carrying across the country for the first time: a glass bell-jar, a wire-gauze cover and the various implements needed for handling and transferring my irascible and dangerous subjects. My search for burrows among the pebbles and the tufts of thyme and lavender is soon successful.

Here is a splendid one. I learn by inserting a straw that it is inhabited by a Tarantula of a size suited to my plans. The soil around the aperture is cleared and flattened to receive the wire-gauze, under which I place a Pompilus. This is the time to light a pipe and wait, lying on the pebbles . . . Yet another disappointment. Half an hour goes by; and the Wasp confines herself to travelling round and round the netting as she did in my study. She gives no sign of greed when confronted with the burrow, though I can see the Tarantula's diamond eyes glittering at the bottom.

The trellised wall is replaced by the glass wall, which, since it does not allow her to scale its heights, will oblige the Wasp to remain on the ground and at last to take cognizance of the shaft, which she seems to ignore. This time we have done the trick!

After a few circuits of her cage, the Calicurgus notices the pit yawning at her feet. She goes down it. This daring confounds me. I should never have ventured to anticipate as much. That she should suddenly fling herself upon the Tarantula when the latter is outside her stronghold, well and good; but to rush into the lair, when the terrible monster is waiting for you below with those two poisoned daggers of hers! What will come of such temerity? A buzzing of wings ascends from the depths. Run to earth in her private apartments, the Lycosa is no doubt at grips with the intruder. That hum of wings is the Calicurgus' paean of triumph, until it be her death-song. The slayer may well be the slain. Which of the two will come up alive?

It is the Lycosa, who hurriedly scampers out and posts herself just over the orifice of the burrow, in her posture of defence, her fangs open, her four front legs uplifted. Can the other have been stabbed? Not at all, for she emerges in her turn, not without receiving on the way a cuff from the Spider, who immediately regains her lair. Dislodged from her basement a second and yet a third time, the Tarantula always comes up unwounded; she always awaits her adversary on her threshold, administers punishment and reenters her dwelling. In vain do I try my two Pompili alternately and change the burrow; I do not succeed in observing anything else. Certain conditions not realized by my stratagems are lacking to complete the tragedy.

Discouraged by the repetition of my futile attempts, I throw up the game, the richer however by one fact of some value: the Calicurgus, without the least fear, descends into the Tarantula's den and dislodges her. I imagine that things happen in the same fashion outside my cages. When expelled from her dwelling, the Spider is more timid and more vulnerable to attack. Moreover, while hampered by a narrow shaft, the operator would not wield her lancet with the precision called for by her designs. The bold irruption shows us once again, more plainly than the tussles on my table, the Lycosa's reluctance to sink her fangs into her enemy's body. When the two are face to face at the bottom of the lair, then or never is the moment to have it out with the foe. The Tarantula is in her own house, with all its conveniences; every nook and corner of the bastion is familiar to her. The intruder's movements are hampered by her ignorance of the premises. Quick, my poor Lycosa, quick, a bite; and it's all up with your persecutor! But you refrain, I know not why, and your reluctance is the saving of the rash invader. The silly Sheep does not reply to the butcher's knife by charging with lowered horns. Can it be that you are the Pompilus' Sheep?

My two subjects are reinstalled in my study under their wire-gauze covers, with bed of sand, reed-stump burrow and fresh honey, complete. Here they find again their first Lycosae, fed upon Locusts. Cohabitation continues for three weeks without other incidents than scuffles and threats which become less frequent day by day. No serious hostility is displayed on either side. At last the Calicurgi die: their day is over. A pitiful end after such an enthusiastic beginning.

Shall I abandon the problem? Why, not a bit of it! I have encountered greater difficulties, but they have never deterred me from a warmly-cherished project. Fortune favours the persevering. She proves as much by offering me, in September, a fortnight after the death of my Tarantula-huntresses, another Calicurgus, captured for the first time. This is the Harlequin Calicurgus (C. scurra, LEP.), who sports the same gaudy costume as the first and is almost of the same size.

Now what does this newcomer, of whom I know nothing, want? A Spider, that is certain; but which? A huntress like this will need a corpulent quarry: perhaps the Silky Epeira (E. serica), perhaps the Banded Epeira (E. fasciata), the largest Spiders in the district, next to the Tarantula. The first of these spreads her large upright net, in which Locusts are caught, from one clump of brushwood to another. I find her in the copses on the neighbouring hills. The second stretches hers across the ditches and the little streams frequented by the dragonflies. I find her near the Aygues, beside the irrigation-canals fed by the torrent. A couple of trips procures me the two Epeirae, whom I offer to my captive next day, both at the same time. It is for her to choose according to her taste.

The choice is soon made: the Banded Epeira is the one preferred. But she does not yield without protest. On the approach of the Wasp, she rises and assumes a defensive attitude, just like that of the Lycosa. The Calicurgus pays no attention to threats: under her harlequin's coat, she is violent in attack and quick on her legs. There is a rapid exchange of fisticuffs; and the Epeira lies overturned on her back. The Pompilus is on top of her, belly to belly, head to head; with her legs she masters the Spider's legs; with her mandibles she grips the cephalothorax. She curves her abdomen, bringing the tip of it beneath her; she draws her sting and...

One moment, reader, if you please. Where is the sting about to strike? From what we have learnt from the other paralysers, it will be driven into the breast, to suppress the movement of the legs. That is your opinion; it was also mine. Well, without blushing too deeply at our common and very excusable error, let us confess that the insect knows better than we do. It knows how to assure success by a preparatory manoeuvre of which you and I had never dreamt. Ah, what a school is that of the animals! Is it not true that, before striking the adversary, you should take care not to get wounded yourself? The Harlequin Pompilus does not disregard this counsel of prudence. The Epeira carries beneath her throat two sharp daggers, with a drop of poison at their points; the Calicurgus is lost if the Spider bites her. Nevertheless, her anaesthetizing demands perfect steadiness of the lancet. What is to be done in the face of this danger which might disconcert the most practised surgeon? The patient must first be disarmed and then operated on.

And in fact the Calicurgus' sting, aimed from back to front, is driven into the Epeira's mouth, with minute precautions and marked persistency. On the instant, the poison-fangs close lifelessly and the formidable quarry is powerless to harm. The Wasp's abdomen then extends its arc and drives the needle behind the fourth pair of legs, on the median line, almost at the junction of the belly and the cephalothorax. At this point the skin is finer and more easily penetrable than elsewhere. The remainder of the thoracic surface is covered with a tough breastplate which the sting would perhaps fail to perforate. The nerve-centres, the source of the leg movements, are situated a little above the wounded point, but the back-to-front direction of the sting makes it possible to reach them. This last wound results in the paralysis of all the eight legs at once.

To enlarge upon it further would detract from the eloquence of this performance. First of all, to safeguard the operator, a stab in the mouth, that point so terribly armed, the most formidable of all; then, to safeguard the larva, a second stab in the nerve-centres of the thorax, to suppress the power of movement. I certainly suspected that the slayers of robust Spiders were endowed with special talents; but I was far from expecting their bold logic, which disarms before it paralyses. So the Tarantula-huntress must behave, who, under my bell-glasses, refused to surrender her secret. I now know what her method is; it has been divulged by a colleague. She throws the terrible Lycosa upon her back, pricks her prickers by stinging her in the mouth and then, in comfort, with a single thrust of the lancet, obtains paralysis of the legs.

I examine the Epeira immediately after the operation and the Tarantula when the Calicurgus is dragging her by one leg to her burrow, at the foot of some wall. For a little while longer, a minute at most, the Epeira convulsively moves her legs. So long as these throes continue, the Pompilus does not release her prey. She seems to watch the progress of the paralysis. With the tips of her mandibles she explores the Spider's mouth several times over, as though to ascertain if the poison-fangs are really innocuous. When all movement subsides, the Pompilus makes ready to drag her prey elsewhere. It is then I take charge of it.

What strikes me more than anything else is the absolute inertia of the fangs, which I tickle with a straw without succeeding in rousing them from their torpor. The palpi, on the other hand, their immediate neighbours, wave at the least touch. The Epeira is placed in safety, in a flask, and undergoes a fresh examination a week later. Irritability has in part returned. Under the stimulus of a straw, I see her legs move a little, especially the lower joints, the tibiae and tarsi. The palpi are even more irritable and mobile. These different movements, however, are lacking in vigour and coordination; and the Spider cannot employ them to turn over, much less to escape. As for the poison-fangs, I stimulate them in vain: I cannot get them to open or even to stir. They are therefore profoundly paralysed and in a special manner. The peculiar insistence of the sting when the mouth was stabbed told me as much in the beginning.

At the end of September, almost a month after the operation, the Epeira is in the same condition, neither dead nor alive: the palpi still quiver when touched with a straw, but nothing else moves. At length, after six or seven weeks' lethargy, real death supervenes, together with its comrade, putrefaction.

The Tarantula of the Ringed Calicurgus, as I take her from the owner at the moment of transportation, presents the same peculiarities. The poison-fangs are no longer irritable when tickled with my straw: a fresh proof, added to those of analogy, to show that the Lycosa, like the Epeira, has been stung in the mouth. The palpi, on the other hand, are and will be for weeks highly irritable and mobile. I wish to emphasise this point, the importance of which will be recognized presently.

I found it impossible to provoke a second attack from my Harlequin Calicurgus: the tedium of captivity did not favour the exercise of her talents. Moreover, the Epeira sometimes had something to do with her refusals; a certain ruse de guerre which was twice employed before my eyes may well have baffled the aggressor. Let me describe the incident, if only to increase our respect a little for these foolish Spiders, who are provided with perfected weapons and do not dare to make use of them against the weaker but bolder assailant.

The Epeira occupies the wall of the wire-gauze cage, with her eight legs widespread upon the trelliswork; the Calicurgus is wheeling round the top of the dome. Seized with panic at the sight of the approaching enemy, the Spider drops to the ground, with her belly upwards and her legs gathered together. The other dashes forward, clasps her round the body, explores her and prepares to sting her in the mouth. But she does not bare her weapon. I see her bending attentively over the poisoned fangs, as though to investigate their terrible mechanism; she then goes away. The Spider is still motionless, so much so that I really believe her dead, paralysed unknown to me, at a moment when I was not looking. I take her from the cage to examine her comfortably. No sooner is she placed on the table than behold, she comes to life again and promptly scampers off! The cunning creature was shamming death beneath the Wasp's stiletto, so artfully that I was taken in. She deceived an enemy more cunning than myself, the Pompilus, who inspected her very closely and took her for a corpse unworthy of her dagger. Perhaps the simple creature, like the Bear in the fable of old, already noticed the smell of high meat.

This ruse, if ruse it be, appears to me more often than not to turn to the disadvantage of the Spider, whether Tarantula, Epeira or another. The Calicurgus who has just put the Spider on her back after a brisk fight knows quite well that her prostrate foe is not dead. The latter, thinking to protect itself, simulates the inertia of a corpse; the assailant profits by this to deliver her most perilous blow, the stab in the mouth. Were the fangs, each tipped with its drop of poison, to open then; were they to snap, to give a desperate bite, the Pompilus would not dare to expose the tip of her abdomen to their deadly scratch. The shamming of death is exactly what enables the huntress to succeed in her dangerous operation. They say, O guileless Epeirae, that the struggle for life has taught you to adopt this inert attitude for purposes of defence. Well, the struggle for life was a very bad counsellor. Trust rather to common sense and learn, by degrees, at your own cost, that to hit back, above all if you can do so promptly, is still the best way to intimidate the enemy.

The remainder of my observations on these insects under glass is little more than a long series of failures. Of two operators on Weevils, one, the Sandy Cerceris (C. arenaria), persistently scorned the victims offered; the other, Ferrero's Cerceris (C. Ferreri), allowed herself to be empted after two days' captivity. Her tactical method, as I expected, is precisely that of the Cleonus-huntress, the Great Cerceris, with whom my investigations commenced. When confronted with the Acorn-weevil, she seizes the insect by the snout, which is immensely long and shaped like a pipe-stem, and plants her sting in its body to the rear of the prothorax, between the first and second pair of legs. It is needless to insist: the spoiler of the Cleoni has taught us enough about this mode of operation and its results.

None of the Bembex-wasps, whether chosen among the huntresses of the Gadfly or among the lovers of the housefly rabble, satisfied my aspirations. Their method is as unknown to me now as at the distant period when I used to watch it in the Bois des Issards. Their impetuous flight, their love of long journeys are incompatible with captivity. Stunned by colliding with the walls of their glass or wire-gauze prison, they all perish within twenty-four hours. Swifter in their movements and apparently satisfied with their honeyed thistle-heads, the Spheges, huntresses of Crickets or Ephippigers, die as quickly of nostalgia. All I offer them leaves them indifferent.

Nor can I get anything out of the Eumenes, notably the biggest of them, the builder of gravel cupolas, Amedeus' Eumenes. All the Pompili, except the Harlequin Calicurgus, refuse my Spiders. The Palarus, who preys upon an indefinite number of the Hymenopteron clan, refuses to tell me if she drinks the honey of the Bees, as does the Philanthus, or if she lets the others go without manipulating them to make them disgorge. The Tachytes do not vouchsafe their Locusts a glance; Stizus ruficornis promptly gives up the ghost, disdaining the Praying Mantis which I provide for her.

What is the use of continuing this list of checks? The rule may be gathered from these few examples: occasional successes and many failures. What can be the reason? With the exception of the Philanthus, tempted from time to time by a bumper of honey, the predatory Wasps do not hunt on their own account; they have their victualling-time, when the egg-laying is imminent, when the family calls for food. Outside these periods, the finest heads of game might well leave these nectar-bibbers indifferent. I am careful therefore, as far as possible, to capture my subjects at the proper season; I give preference to mothers caught upon the threshold of the burrow with their prey between their legs. This diligence of mine by no means always succeeds. There are demoralized insects which, once under glass, even after a brief delay, no longer care about the equivalent of their prize.

All the species do not perhaps pursue their game with the same ardour; mood and temperament are more variable even than conformation. To these factors, which are of the nicest order, we may add that of the hour, which is often unfavourable when the subject is caught at haphazard on the flowers, and we shall have more than enough to explain the frequency of the failures. After all, I must beware of representing my failures as the rule: what does not succeed one day may very well succeed another day, under different conditions. With perseverance and a little skill, any one who cares to continue these interesting studies will, I am sure, fill up many gaps. The problem is difficult but not impossible.

I will not quit my bell-jars without saying a word on the entomological tact of the captives when they decide to attack. One of the pluckiest of my subjects, the Hairy Ammophila, was not always provided with the hereditary dish of her family, the Grey Worm. I offered her indiscriminately any bare-skinned caterpillars that I chanced to find. Some were yellow, some green, some brown with white edges. All were accepted without hesitation, provided that they were of suitable size. Tasty game was recognized wonderfully under very dissimilar liveries. But a young Zeuzera-caterpillar, dug out of the branches of a lilac-tree, and a silkworm of small dimensions were definitely refused. The overfed products of our silkworm-nurseries and the mystery-loving caterpillar which gnaws the inner wood of the lilac inspired her with suspicion and disgust, despite their bare skin, which favoured the sting, and their shape, which was similar to that of the victims accepted.

Another ardent huntress, the Interrupted Scolia, refused the Cetonia-grub, which is of like habits with the Anoxia-larva; the Two-banded Scolia also refused the Anoxia. The Philanthus, the headlong murderess of Bees, saw through my trickery when I confronted her with the Virgilian Bee, the Eristalis (E. tenax). She, a Philanthus, take this Fly for a Bee! What next! The popular idea is mistaken; antiquity too is mistaken, as witness the "Georgics," which make the putrid remains of a sacrificed Bull give birth to a swarm; but the Wasp makes no mistake. In her eyes, which see farther than ours, the Eristalis is an odious Dipteron, a lover of corruption, and nothing more.

CHAPTER 14.

OBJECTIONS AND REJOINDERS.

***

No idea of any scope can begin its soaring flight but straightway the curmudgeons are after it, eager to break its wings and to stamp the wounded thing under foot. My discovery of the surgical methods that give the Hunting Wasps their preserved foodstuffs has undergone the common rule. Let theories be discussed, by all means: the realm of the imagination is an untilled domain, in which every one is free to plant his own conceptions. But realities are not open to discussion. It is a bad policy to deny facts with no more authority than one's wish to find them untrue. No one that I know of has impugned by contrary observations what I have so long been saying about the anatomical instinct of the Wasps that hunt their prey; instead, I am met with arguments. Mercy on us! First use your eyes and then you shall have leave to argue! And, to persuade people to use their eyes, I mean to reply, since we have time to spare, to the objections which have been or may be raised. Of course, I pass over in silence those in which childish disparagement shows its nose too plainly.

The sting, I am told, is directed at one point rather than another because that is the only vulnerable point. The insect cannot choose what wound it will inflict; it stings where it must. Its wonderful operative method is the necessary result of the victim's structure. Let us first, if we attach any importance to lucidity, come to an understanding about the word "vulnerable." Do you mean by this that the point or rather points wounded by the sting are the only points at which a lesion will suddenly cause either death or paralysis? If so, I share your opinion; not only do I share it, but I was the first to proclaim it. My whole thesis is contained in that. Yes, a hundred times yes, the points wounded are the only vulnerable points; they are even very vulnerable; they are the only points which lend themselves to the infliction of sudden death or else paralysis, according to the operator's intention.

But this is not how you understand the matter: you mean accessible to the sting, in a word, penetrable. Here we part company. I have against me, I admit, the Weevils and the Buprestes of the Cerceres. These mailed ones hardly give the sting a chance, save behind the prothorax, the point at which the lancet is actually directed. If I were one to stand on trifles, I might observe that in front of the prothorax, under the throat, is an accessible spot and that the Cerceres will have nothing to do with it. But let us proceed; I give up the horn-clad Beetle.

What are we to say of the Grey Worm and other caterpillars beloved of the Ammophilae? Here are victims accessible to the sting underneath, on the back, on the sides, fore and aft, everywhere with the same facility, excepting the top of the head. And of this infinity of points, which are equally penetrable, the Wasp selects ten, always the same, differing in no way from the rest, unless it be by the close proximity of the nerve-centres. What are we to say of the Cetonia - and Anoxia-larvae, which are always attacked in the first thoracic segment, after long and painful struggles, when the assailant can sting the grub freely at whatever point she chooses, since it is quite naked and offers no greater resistance to the lancet at one point than at another?

What are we to think of the Sphex' Crickets and Ephippigers, stabbed three times on the side of the thorax, which is fairly well defended, whereas the abdomen, soft and bulky, into which the sting would sink like a needle into a pat of butter, is neglected? Do not let us forget the Philanthus, who takes no account either of the fissures beneath the abdominal plates or of the wide hiatus behind the corselet, but plunges her weapon, at the base of the throat, through a gap of a fraction of a millimetre. Let us just mention the Mantis-hunting Tachytes. Does she make for the most undefended point when she stabs, first of all, at its base, the Mantis' dreadful engine - the arm-pieces each fitted with a double saw - at the risk of being seized, transfixed and crunched on the spot if she misses her blow? Why does she not strike at the creature's long abdomen? That would be quite easy and free from danger.

And the Calicurgi, if you please. Are they also unskilled duelists, plunging the dirk into the only easily accessible point, when their very first move is to paralyse the poison-fangs? If there is one point about the Tarantula and the Epeira that is dangerous and difficult to attack, it is certainly the mouth which bites with its two poisoned harpoons. And these desperadoes dare to brave that deadly trap! Why do they not follow your judicious advice? They should sting the plump belly, which is wholly unprotected. They do not; and they have their reasons, as have the others.

All, from the first to the last, show us, clear as water from the rock, that the outer structure of the victims operated on counts for nothing in the method of operating. This is determined by the inner anatomy. The points wounded are not stung because they are the only points penetrable by the lancet; they are stung because they fulfil an important condition, without which penetrability loses its value. This condition is none other than the immediate proximity of the nerve-centres whose influence has to be suppressed. When at close quarters with her prey, whether soft or armour-clad, the huntress behaves as if she understood the nervous system better than any of us. The thoughtless objection about the only penetrable points is, I hope, swept aside forever.

I am also told:

"It is possible, if it comes to that, for the sting to be delivered in the neighbourhood of the nerve-centres; in a victim at most three or four centimetres long, distances are very small. But a casual there or thereabouts is a very different thing from the precision of which you speak."

Oh, they are "thereabouts," are they? We shall see! You want figures, millimetres, fractions? You shall have them!

First I call to witness the Interrupted Scolia. If the reader no longer has her method of operating in mind, I will beg him to refresh his memory. The two adversaries, in the preliminary conflict, may be fairly well represented by two rings interlocked not in the same plane but at right angles. The Scolia grips a point of the Anoxia-grub's thorax; she curves her body underneath it and, while encircling the grub, gropes with the tip of her abdomen along the median line of the larva's neck. Owing to her transversal position, the assailant is now free to aim her weapon in a slightly slanting direction, whether towards the head or towards the thorax, at the same point of entry in the larva's throat. Between the two opposite slants of the sting, which is itself very short, what can the distance be? Two millimetres, perhaps less. That is very little. No matter: let the operator make a mistake of this length – negligible, you may tell me – let the sting slant towards the head instead of slanting towards the thorax; and the result of the operation will be entirely different. With a slant towards the head, the cerebral ganglia are wounded and their lesion causes sudden death. This is the stroke of the Philanthus, who kills her Bee by stinging her from below, under the chin. The Scolia needed a motionless but not dead victim, one that would supply fresh victuals; she will now have only a corpse, which will soon go bad and poison the larva.

With a slant towards the thorax, the sting wounds the little mass of nerve-cells in the thorax. This is the regulation stroke, the one which will induce paralysis and leave the small amount of life needed to keep the provisions fresh. A millimetre higher kills; a millimetre lower paralyses. On this tiny deviation the salvation of the Scolia race depends. You need not fear that the operator will make any mistake in this micrometrical performance: her sting always slants towards the thorax, although the opposite inclination is just as practicable and easy. What would be the outcome of a there or thereabouts under these conditions? Very often a corpse, a form of food fatal to the grub.

The Two-banded Scolia stings a little lower down, on the line of demarcation between the first two thoracic segments. Her position is likewise transversal in relation to the Cetonia-grub; but the distance of the cervical ganglia from the point where the sting enters would possibly not allow the weapon turned towards the head to inflict a lesion followed by sudden death as in the above instance. I am calling this witness with another object. It is extremely unusual for the operator, no matter what her prey or her method, to make a slight mistake and sting merely somewhere near the requisite point. I see them all groping with the tip of the abdomen, sometimes seeking persistently, before unsheathing. They thrust only when the point beneath the sting is precisely that at which the wound will produce its full effect. The Two-banded Scolia in particular will struggle with the Cetonia-grub for half an hour at a time to enable herself to drive in the stiletto at the right spot.

Wearied by an endless scuffle, one of my captives committed before my eyes a slight blunder, an unprecedented thing. Her weapon entered a little to one side, not quite a millimetre from the central point and still, of course, on the line of demarcation between the first two thoracic segments. I at once laid hold of the precious specimen, which was to teach me curious matters about the effects of an ill-delivered stroke. If I myself had made the insect sting at this or that point, there would have been no particular interest in it: the Scolia, held between the fingertips, would wound at random, like a Bee defending herself; her undirected sting would inject the poison at haphazard. But here everything happened by rule, except for the little error of position.

Well, the victim of this clumsy operation has its legs paralysed only on the left side, the side towards which the weapon was deflected; it is a case of hemiplegia.

The legs on the right side move. If the operation had been performed in the normal fashion the result would have been sudden inertia of all six legs. The hemiplegia, it is true does not last long. The torpor of the left half rapidly gains the right half of the body and the creature lies motionless, incapable of burying itself in the mould, without, however, realizing the conditions indispensable to the safety of the egg or the young grub. If I seize one of its legs or a point of the skin with the tweezers, it suddenly shrivels and curls up and swells out again, as it does when in complete possession of its energies. What would become of an egg laid on such victuals? At the first closing of this ruthless vice, at the first contraction, it would be crushed, or at least detached from its place; and any egg removed from the point where the mother has fastened it is bound to perish. It needs, on the Cetonia's abdomen, a yielding support which the bites of the newborn larva will not set aquiver. The slightly eccentric sting gives none of this soft mass of fat, always outstretched and quiescent. Only on the following day, after the torpor has made progress, does the larva become suitably inert and limp. But it is too late; and in the meantime the egg would be in serious danger on this half-paralysed victim. The sting, by straying less than a millimetre, would leave the Scolia without progeny.

I promised fractions. Here they are. Let us consider the Tarantula and the Epeira on whom the Calicurgi have just operated. The first thrust of the sting is delivered in the mouth. In both victims the poison-fangs are absolutely lifeless: tickling with a bit of straw never once succeeds in making them open. On the other hand, the palpi, their very near neighbours, their adjuncts as it were, possess their customary mobility. Without any previous touches, they keep on moving for weeks. In entering the mouth the sting did not reach the cervical ganglia, or sudden death would have ensued and we should have before our eyes corpses which would go bad in a few days, instead of fresh carcases in which traces of life remain manifest for a long time. The cephalic nerve-centres have been spared.

What is wounded then, to procure this profound inertia of the poison-fangs? I regret that my anatomical knowledge leaves me undecided on this point. Are the fangs actuated by a special ganglion? Are they actuated by fibres issuing from centres exercising further functions? I leave to anatomists equipped with more delicate instruments than I the task of elucidating this obscure question. The second conjecture appears to me the more probable, because of the palpi, whose nerves, it seems to me, must have the same origin as those of the fangs. Basing our argument on this latter hypothesis, we see that the Calicurgus has only one means of suppressing the movement of the poisoned pincers without affecting the mobility of the palpi, above all without injuring the cephalic centres and thus producing death, namely, to reach with her sting the two fibres actuating the fangs, fibres as fine as a hair.

I insist upon this point. Despite their extreme delicacy, these two filaments must be injured directly; for, if it were enough for the sting to inject its poison "there or thereabouts," the nerves of the palpi, so close to the first, would undergo the same intoxication as the adjacent region and would leave those appendages motionless. The palpi move; they retain their mobility for a considerable period; the action of the poison, therefore, is evidently situated in the nerves of the fangs.

There are two of these nerve-filaments, very fine, very difficult to discover, even by the professional anatomist. The Calicurgus has to reach them one after the other, to moisten them with her poison, possibly to transfix them, in any case to operate upon them in a very restricted manner; so that the diffusion of the virus may not involve the adjoining parts. The extreme delicacy of this surgery explains why the weapon remains in the mouth so long; the point of the sting is seeking and eventually finds the tiny fraction of a millimetre where the poison is to act. This is what we learn from the movements of the palpi close to the motionless fangs; they tell us that the Calicurgi are vivisectors of alarming accuracy.

If we accept the hypothesis of a special nerve-centre for the mandibles, the difficulty would be a little less, without detracting from the operator's talent. The sting would then have to reach a barely visible speck, an atom in which we should hardly find room for the point of a needle. This is the difficulty which the various paralysers solve in ordinary practice. Do they actually wound with their dirks the ganglion whose influence is to be done away with? It is possible, but I have tried no test to make sure, the infinitely tiny wound appearing to be too difficult to detect with the optical instruments at my disposal. Do they confine themselves to lodging their drop of poison on the ganglion, or at all events in its immediate neighbourhood? I do not say no.

I declare moreover, that, to provoke lightning paralysis, the poison, if it is not deposited inside the mass of nervous substance, must act from somewhere very near. This assertion is merely echoing what the Two-banded Scolia has just shown us: her Cetonia-grub, stung less than a millimetre from the regular spot, did not become motionless until next day. There is no doubt, judging by this instance, that the effect of the virus spreads in all directions within a radius of some extent; but this diffusion is not enough for the operator, who requires for her egg, which is soon to be laid, absolute safety from the very first.

On the other hand, the actions of the paralysers argue a precise search for the ganglia, at all events for the first thoracic ganglion, the most important of all. The Hairy Ammophila, among others, affords us an excellent example of this method. Her three thrusts in the caterpillar's thorax and especially the last, between the first and second pair of legs, are more prolonged than the stabs distributed among the abdominal ganglia. Everything justifies us in believing that, for these decisive inoculations, the sting seeks out the corresponding ganglion and acts only when it finds it under its point. On the abdomen this peculiar insistence ceases; the sting passes swiftly from one segment to another. For these segments, which are less dangerous, the Ammophila perhaps relies on the diffusion of her venom; in any case, the injections, though hastily administered, do not diverge from a close vicinity of the ganglia, for their field of action is very limited, as is proved by the number of inoculations necessary to induce complete torpor, or, more simply, by the following example.

A Grey Worm which had just received its first sting on the third thoracic segment repulses the Ammophila and with a jerk hurls her to a distance. I profit by the occasion and take hold of the grub. The legs of this third segment only are paralysed; the others retain their usual mobility. However helpless in the two injured legs, the animal can walk very well; it buries itself in the earth, returning to the surface at night to gnaw the stump of lettuce with which I have served it. For a fortnight my paralytic retains perfect liberty of action, except in the segment operated on; then it dies, not of its wound but accidentally. All this time the effect of the poison has not spread beyond the inoculated segment.

At any point where the sting enters, anatomy informs us of the presence of a nervous nucleus. Is this centre directly smitten by the weapon? Or is it poisoned with virus, from a very small distance, by the progressive impregnation of the neighbouring tissues? This is the doubtful point, though it does not in any way invalidate the precision of the abdominal injections, which are comparatively neglected. As for those in the caterpillar's thorax, their precision is beyond dispute. After the Ammophilae, the Scoliae and, above all, the Calicurgi, is it really necessary to bring into court yet other witnesses, who would all swear that, with modifications of detail, the movement of their lancet is strictly regulated by the nervous system of the prey? This ought to be enough. The proof is established for those who have ears to hear with.

Others delight in objections whose oddity surprises me. They see in the poison of the Hunting Wasps an antiseptic liquid and in victuals stored in their burrows preserved meats which are kept fresh not by a remnant of life but by the virus and its microbes. Come, my learned masters, let us just talk the matter over, between ourselves. Have you ever seen the larder of a skilled Hunting Wasp, a Sphex for instance, a Scolia, an Ammophila? You haven't, have you? I thought as much. Yet it would be better to begin by doing so, before bringing the preservative microbe on the scene. The slightest examination would have shown you that the victuals cannot be compared exactly with smoked hams. The thing moves, therefore it is not dead. There you have the whole matter, in its artless simplicity. The palpi move, the mandibles open and shut, the tarsi quiver, the antennae and the abdominal filaments wave to and fro, the abdomen throbs, the intestine rejects its contents, the animal reacts to the stimulus of a needle, all of which signs are hardly compatible with the idea of pickled meat.

Have you had the curiosity to look through the pages in which I set forth the detailed results of my observations? You haven't, have you? Again, I thought as much. It is a pity. You would there find, in particular, the history of certain Ephippigers who, after being stung by the Sphex according to rule, were reared by myself by hand. You must agree that these are queer preserves to be produced by the use of an antiseptic fluid. They accept the mouthfuls which I offer them on the tip of a straw; they feed, they sit up and take nourishment. I shall never live to see tinned sardines doing as much.

I will avoid tedious repetition and content myself with adding to my old sheaf of proofs a few facts which have not yet been related. The Nest-building Odynerus showed us in her cells a few Chrysomela-larvae fixed by the hinder part to the side of the reed. The grub fastens itself in this way to the poplar-leaf to obtain a purchase when the moment has come for leaving the larval slough. Do not these preparations for the nymphosis tell us plainly that the creature is not dead?

The Hairy Ammophila affords us an even better example. A number of caterpillars operated on before my eyes attained, some sooner, some later, the chrysalis stage. My notes are explicit on the subject of some of them, taken on Verbascum sinuatum. Sacrificed on the 14th of April, they were still irritable when tickled with a straw a fortnight after. A little later, the pale-green colouring of the early stages is replaced by a reddish brown, except on two or three segments of the median ventral surface. The skin wrinkles and splits, but does not come detached of its own accord. I can easily remove it in shreds. Under this slough appears the firm, chestnut-brown horn integument of the chrysalis. The development of the nymphosis is so correct that for a moment the crazy hope occurs to me that I may see a Turnip-moth come out of this mummy, the victim of a dozen dagger-thrusts. For the rest, there is no attempt at spinning a cocoon, no jet of silky threads flung out by the caterpillar before turning into a chrysalis. Perhaps under normal conditions metamorphosis takes place without this protection. However, the moth whom I expected to see was beyond the limits of the possible. In the middle of May, a month after the operation on the caterpillars, my three chrysalids, still incomplete underneath, in the three or four middle segments, withered and at last went mouldy. Is the evidence conclusive this time? Who can conceive such a silly idea as that a prey really dead, a corpse preserved from putrefaction by an antiseptic, could contain what is perhaps the most delicate work of life, the development of the grub into the perfect insect?

The truth must be driven into recalcitrant brains with great blows of the sledgehammer. Let us once more employ this method. In September I unearth from a heap of mould five Cetonia-grubs, paralysed by the Two-banded Scolia and bearing on the abdomen the as yet unhatched egg of the Wasp. I remove the eggs and install the helpless creatures on a bed of leaf-mould with a glass cover. I propose to see how long I can keep them fresh, able to move their mandibles and palpi. Already the victims of various Hunting Wasps had instructed me on a similar matter; I knew that traces of life linger for two, three, four weeks and longer. For instance, I had seen the Ephippigers of the Languedocian Sphex continue the waving of their antennae and their paralytic shudders for forty days of artificial feeding by hand; and I used to wonder whether the more or less early death of the other victims was not due to lack of nourishment quite as much as to the operation which they had undergone. However, the insect in its adult form usually has a very brief existence. It soon dies, killed by the mere fact of living, without any other accident. A larva is preferable for these investigations. Its constitution is livelier, better able to support protracted abstinence, above all during the winter torpor. The Cetonia-grub, a regular lump of bacon, nourished by its own fat during the winter season, fulfils the needful conditions to perfection. What will become of it, lying belly upwards on its bed of leaf-mould? Will it survive the winter?

At the end of a month, three of my grubs turn brown and lapse into rottenness. The other two keep perfectly fresh and move their antennae and palpi at the touch of a straw. The cold weather comes and tickling no longer elicits these signs of life. The inertia is complete; nevertheless their appearance remains excellent, without a trace of the brownish tinge, the sign of deterioration. At the return of the warm weather, in the middle of May, there is a sort of resurrection. I find my two larvae turned over, belly downwards; much more: they are half-buried in the mould. When teased, they coil up lazily; they move their legs as well as their mouthparts, but slowly and without vigour. Then their strength seems to revive. The convalescent, resuscitated grubs dig with clumsy efforts into their bed of mould; they dive into it and disappear to a depth of about two inches. Recovery seems to be imminent.

I am mistaken. In June I unearth the invalids. This time, the larvae are dead; their brown colour tells me as much. I expected better things. Never mind: this is no trifling success. For nine months, nine long months, the grubs stabbed by the Scolia kept fresh and alive. Towards the end, torpor was dispelled, strength and movement returned, sufficiently to enable them to leave the surface where I had placed them and to regain the depths by boring a passage through the soil. I really think that after this resurrection there will be no more talk of antiseptics, unless and until tinned Herrings begin to frolic in their brine.

INDEX

www.ReadHowYouWant.com

You can buy our **Large Type** and **EasyRead** books from our www.ReadHowYouWant.com website, from websites like Amazon.com and through your UK and North American bookshop.

EasyRead books are designed to make your reading easy and enjoyable. **EasyRead** books are published in different font sizes, so you can select the font size best for you.

EasyRead is for people with normal eyesight who want books in an easy-to-read format.

EasyRead Comfort is for people who find reading small print tiring but do not need large print.

EasyRead Large is for people who find it easier to read larger print.

The **EasyRead** font, character, word and line spacing have all been set to make it as easy as possible both to recognize a word, and to run your eye along a line of text without losing your place. We split as few words as possible at line ends, as split words make reading harder and can be annoying. For out-of-copyright books, words have been changed to modern spelling (where the sense is not affected), and the original hyphenation of compound words has been retained where this improves word recognition or sense.

With most things you buy, you get a choice, and you can choose the make and model that suits best. There is a big exception – books. You don't get to choose a format that is easy for you to read – you have to read the format the publisher selects.

This **"One Size Fits All"** paradigm **no longer** need apply to books.

At www.ReadHowYouWant.com, you can order a book just as you want it, so you can read it easily. You can choose nearly any format, and at a surprisingly low cost, because of a technical breakthrough that allows us to typeset an individual book automatically, print and bind the book and have it quickly sent directly to you.

You can have your book in **Talking Book** formats (**DAISY or MP3**) or as an **E-book** or in **Braille**.

We have developed totally new visual formats which we hope will help people with **dyslexia** and other print reading disabilities. Look at www.ReadHowYouWant.com for more information.

Good news for publishers and authors. There is a big market for large type, personalized and accessible format books. We can help publishers and authors find new market opportunities for their books, and selling your book through www.ReadHowYouWant.com is easy – we do the work for you. Contact us on info@ReadHowYouWant.com.

Made in the USA
Monee, IL
04 November 2026

67383589R00154